g
nature
a home

Ladybirds

Richard Comont

B I F E
LC SYDNEY

BLOOMSBURY WILDLIFE
Bloomsbury Publishing Plc
50 Bedford Square, London, WC1B 3DP, UK

BLOOMSBURY, BLOOMSBURY WILDLIFE and the Diana logo are trademarks of Bloomsbury Publishing Plc

First published in the United Kingdom, 2019

A catalogue record for this book is available from the British Library

Library of Congress Cataloguing-in-Publication data has been applied for.

ISBN: PB: 978-1-4729-5585-2; ePDF: 978-1-4729-5584-5; ePub: 978-1-4729-5586-9

2 4 6 8 10 9 7 5 3 1

Design by Susan McIntyre
Printed and bound in India by Replika Press Pvt. Ltd.

MIX
Paper from responsible sources
FSC
www.fsc.org
FSC® C016779

To find out more about our a … for our newsletters

For all items sold, Blooms … blisher's receipts from sales of licensed titles … ubsequent sellers of this book are not com … rities Act 1992.

Contents

Meet the Ladybirds

Everyone loves a ladybird. Bright red, spattered with black spots, and clumsily endearing as they amble around the garden, ladybirds have been seen as lucky symbols of summer through the centuries. Farmers and gardeners appreciate them as guardians of plants against the greenfly hordes, and historically ladybirds have been connected to God or the Virgin Mary in thanks for the role they play in crop protection. Often associated with children, they have also been immortalised in rhymes and stories. But beyond our mental image of the lucky ladybird lies a hugely successful range of species, far more diverse than we give them credit for.

Ladybirds – either the real thing or their images – are never far away. Live ladybirds crowd our gardens and parks, trees and roadside verges, heathland, moorland and pond edges. Some species even venture inside our houses during the winter; once there, they are likely to meet pictures of themselves on books, clothing or any one of the huge number of items produced by organisations that use ladybirds as logos. One species in particular has a strong claim to be the single most iconic species of insect worldwide: the Seven-spot Ladybird (*Coccinella septempunctata*) is the symbol of Ladybird clothing,

Opposite: Ladybirds are familiar to most people, and can often be found in parks and gardens.

Left: The Seven-spot Ladybird is perhaps the most iconic of all the ladybird species.

Above: Some ladybirds are familiar because they share our homes in winter, such as these Harlequin Ladybirds.

Below: Farmers and gardeners value ladybirds for their pest-eating habits, which ultimately help protect crops and plants.

Dutch street tiles and much else besides, as well as the state insect of five US states and the star of more postage stamps than any other insect.

Despite the 'lady-' prefix to their common name, ladybirds are not all female, and nor are they birds (or bugs, despite the American name 'ladybug'); instead, they are beetles. Ladybirds make up just over 1 per cent of the 4,000-plus British beetle species, but, brightly coloured and abundant in gardens, they punch far above their weight in terms of familiarity.

A diverse group

Not only are ladybirds common and widespread, they also come in countless colours and sizes. The Seven-spot Ladybird is perhaps the most iconic British insect, and the other black-spotted red species are swept along on its coat-tails of familiarity. But it is just one of 47 ladybird species resident in the UK and more than 6,000 species worldwide, found on every continent except Antarctica.

Compared with many other countries, the UK's total of 47 ladybird species is low; in contrast, the USA has 481 known species and Canada 162. The English Channel has acted as a barrier since the end of the last ice age, almost certainly preventing the colonisation of Britain by a range of species that are widespread on the north coasts of France, Belgium and Holland, including *Oenopia conglobata*, *Cynegetis impunctata* and *Calvia decemguttata*. However, even the British ladybird fauna is surprisingly diverse, and representative of the group across the world. It contains both the smallest and the largest species in Europe – the 1mm (1/16in) Dot Ladybird (*Stethorus pusillus*) and the 9mm (5/16in) Eyed Ladybird (*Anatis ocellata*), respectively – alongside a range of predators, herbivores and fungivores (fungus-feeders).

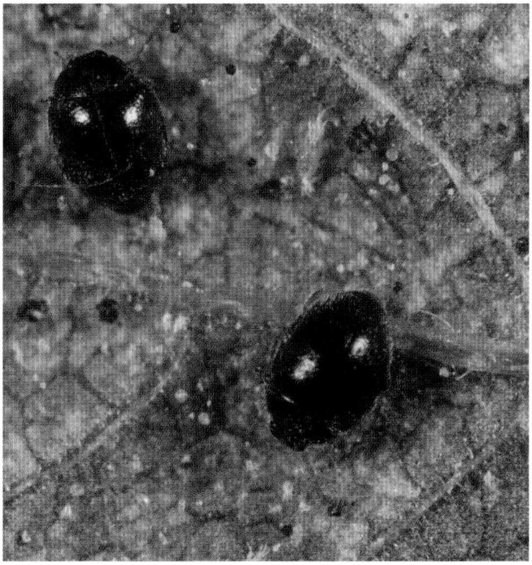

Below: Ladybirds come in a range of different sizes: from Britain's smallest ladybird, the Dot Ladybird (left), to our largest species, the Eyed Ladybird (right), which eats aphids that are bigger than its tiny cousin.

Right: The 18-spot Ladybird is easily recognised by the cream markings smeared across a maroon background.

Worldwide, the smallest ladybirds are a mere 0.6mm ($^3/_{32}$in) long – not much bigger than a full stop – and even the largest are hardly giants at just more than 1cm ($^3/_8$in). Most are spotty, though not just black on red: the background can be red, orange, yellow, brown, black or another colour, while the spots can be black, purple, white, red, orange or yellow. And the spots aren't just round either – they may be oval, triangular or kidney-shaped, or streaks, bullseye roundels or indeterminate blobs. Many ladybird species are brightly coloured but some are camouflaged – a few species even manage to be both at the same time!

Below: Pine Ladybirds are distinguished by their black colour, with oval and comma-shaped red markings.

Recognising ladybirds

The UK's 47 resident ladybirds include 26 'true' or 'conspicuous' ladybirds – generally the larger (3–10mm, or 1/8–3/8in), brightly coloured, hairless species that are immediately recognisable as 'ladybirds' and (in the UK at least) have common names – and 21 'inconspicuous' ladybirds, which are generally smaller and hairier, with much duller colouration and usually only scientific names. Both groups are occasionally reinforced through immigration – around 60 ladybird species have been found in Britain in total, with many of the additional species accidentally imported on fruit and vegetables.

Ladybirds are usually pretty distinctive, but even among Britain's 26 conspicuous ladybird species there is a lot of variation. Most species are orange or red and black, but this can be either red with black spots (like the Seven-spot), or black with red spots (like the Kidney-spot Ladybird, *Chilocorus renipustulatus*). Some species fluctuate between the two: the Two-spot (*Adalia bipunctata*), 10-spot (*A. decempunctata*) and Harlequin (*Harmonia axyridis*) Ladybirds each have a range of colour forms, including some that are red with two to 21 black spots, and others that are black with two to six red spots. Three species are

Below: British ladybird species vary massively in colour, but their shape and general appearance remain similar.

Above: The unique markings of ladybirds can vary a great deal, from the neat spots of a 22-spot Ladybird (left) to the smears and streaks on a Striped Ladybird (right).

yellow with black spots, and four are orange or brown with white markings (although these vary between spots and stripes). The Eyed Ladybird has both: black spots, ringed with beige, over an orangey-red background.

High-vis beetles

Many people don't realise that ladybirds are beetles until the fact is pointed out to them. If you look past the colours, they are actually entirely typical for the group, with six legs, three body sections, one pair of antennae, biting mouthparts and hardened front wings formed into elytra (wing cases). All ladybird species fall within the family Coccinellidae, which is split into a range of subfamilies, as well as the less scientific split between the conspicuous and inconspicuous species. The family name Coccinellidae comes either from the Greek *kokkos*, meaning 'berry', or *coccinatus*, meaning 'scarlet', both of which fit the typical ladybird species well.

Everyone can recognise a typical ladybird, but it becomes much trickier to pick out the less standard species. Not all are bright red; indeed, some are not brightly coloured at all. Because of this diversity, the ladybirds are surprisingly difficult to pin down as a family. Generally, they can be recognised by their bright colours, spotted markings and hemispherical dome-like shape, and their ability to produce brightly coloured, foul-tasting

Left: Ladybirds, like this Seven-spot, are clearly dome-shaped when viewed from the front or from the side.

defensive secretions, but all these rules of thumb have exceptions. Confirming some of the more unusual species as ladybirds can require close examination, particularly of mouthparts (ladybirds generally have large triangular, axe-like palps) and legs (to count the number of tarsal segments – a ladybird's 'toes').

Left: In addition to their recognisable domed shape, some ladybirds are also slightly hairy, as seen here in this 24-spot Ladybird.

Anatomy

Like all insects, ladybirds have three body sections: a head, thorax and abdomen. The head is at the front and carries the sensory and feeding apparatus. Each ladybird has two antennae, which are relatively short and slightly clubbed, and two large compound eyes. The mouthparts are adapted for biting, with rounded, sickle-shaped mandibles that differ in shape according to the preferred prey type of the ladybird species. Below and behind the mandibles are the labial and maxillary palps. These are short antenna-like appendages, with axe-shaped final segments, and are used for manipulating and tasting food before it enters the mouth.

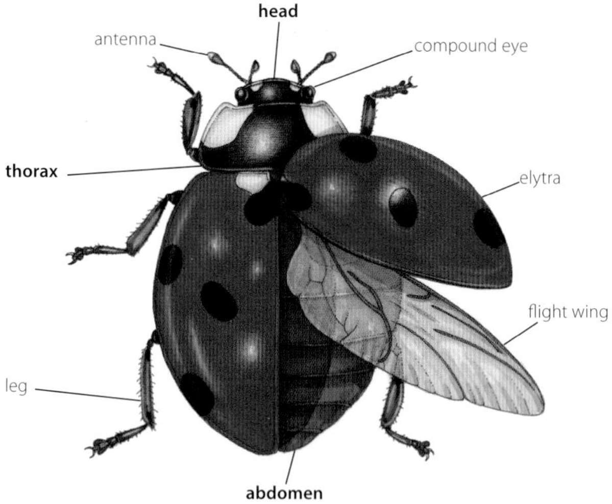

The abdomen is at the rear and is mainly filled with the ladybird's digestive and reproductive organs, as well as the fat-storage areas. Arched over the top like a domed tent are the elytra – these wing cases are the brightly coloured external parts of the ladybird that are very noticeable. They are actually the ladybird's front pair of wings, hardened to act as a protective casing. In most ladybirds they cover the gossamer-thin flight wings, which sit folded up beneath them, on top of the abdomen. However, a few species do not have flight wings beneath their elytra and cannot fly. The two sets of wings – flight wings and elytra – are actually made of almost exactly the same substance, chitin, just laid down in very different thicknesses.

The wings are attached to the body at the top of the thorax. In ladybirds, this section of the body is short but broad, tucked between the head and the abdomen, and filled with muscles to power the legs and wings. The legs are short and can be folded into grooves on the underside of the body for protection. The elytra (and the rest of the body) can sometimes be covered in short, downy hairs, although this tends to be restricted to the smaller species in the UK, which may not be immediately recognisable as ladybirds.

Markings

A hypothetical 'typical ladybird' would be brightly coloured with numerous black spots, to warn off predators. Both the base colour and spot colour and number can vary enormously between species, and often within species as well. Although spot numbers are often used in species names – as in the Seven-spot Ladybird, which has the specific epithet *septempunctata* – the variation within species means that simply counting spots is not a reliable method for identification! And despite the myth, nor is it true that ladybirds have a spot for each age year, as they live only to a maximum of one year.

A few species even change colour through their adult lifetime. For instance, the North American 15-spot Ladybug (*Anatis labiculata*) changes from pale grey with distinct black spots to a dark purple-brown with spots that are no longer visible against the background.

Below: These are both Harlequin Ladybirds, which proves that there is more to identification than just counting spots!

Feeding and life cycle

For ladybirds generally, home is where the food is. Different species do show preferences for different habitat types, but the main factor determining where they can be found is the availability of something tasty to snack on. Unsurprisingly for a group with this level of devotion to eating, ladybirds are pretty voracious. Indeed, the Seven-spot Ladybird is renowned for eating around 5,000 aphids in its lifetime. While aphids (including greenfly and blackfly) form the bulk of the diet of most UK ladybirds, there is a surprising diversity of menu options across species. A significant minority prefer to eat scale insects (small, flattened insects that suck plant sap), and a few graze on mildew growing on leaves. Others make a dash for the nearest whitefly or spider mite, while one species seems to live mainly on the larvae of other beetles. Two British species are even herbivorous, leaving pale windows scraped in the leaves of False Oat-grass (*Arrhenatherum elatius*) or White Bryony (*Bryonia alba*). Worldwide, some herbivorous species can reach such numbers that they occasionally become pests; conversely, many of the

Below: Ladybird eggs are very similar in appearance, making it difficult to identify the species from the egg. The elongated yellow eggs usually take 4–5 days to hatch into larvae.

Above: Batches of ladybird eggs soon give rise to many tiny, predatory larvae. Ladybird larvae (left) will devour lots of aphids until they are ready to pupate (right).

scale-insect predators play a vital role in the control of crop pests, particularly in orchards.

In all these species, the fully grown larva is the most voracious stage. Ladybirds have a four-stage life cycle, as seen in butterflies and moths, and the larvae are the equivalent of the caterpillar stage. Like caterpillars, they exist for just one thing – to eat – and are very different to the smooth, rounded adults. Almost crocodilian in appearance, they are long, tubular eating machines. Often studded with warts, hairs or spines, ladybird larvae are long-legged and surprisingly fast-moving.

Larvae are mostly found from late spring through to late summer, generally around aphids or other food sources. They hatch from upright rugby-ball-shaped yellow eggs, and in under a month grow from less than 1mm ($^1/_{16}$in) long to (in some cases) more than 10mm ($^3/_8$in). Once fully grown, the larvae pupate and spend a week or two as a pupa (the equivalent of a butterfly's chrysalis), before hatching into an adult ladybird. This usually happens in mid- to late summer, giving the adult ladybird a little more than a month to eat as much as possible to see it through its winter dormancy.

When spring arrives the following year, the surviving adult ladybirds emerge. They disperse to find food and mates, and the whole cycle begins again.

Where to find ladybirds

A few ladybirds are almost completely ubiquitous in Britain. Although the archetypal placement of a Seven-spot Ladybird is in a garden, in fact the species can be found from Land's End to John o'Groats, and from grassland to conifer plantations. And nor are the 14-spot (*Propylea quattuordecimpunctata*) and Two-spot (or, nowadays, the Harlequin) likely to be far away. Other species are similarly widespread but prefer to live in particular (fairly broad) habitat types. For example, different species are specialised for living in conifers or deciduous trees, moorland or grassland, and even in ant nests or tall vegetation growing out of ponds. The 10-spot and Cream-spot (*Calvia quattuordecimguttata*) Ladybirds like deciduous trees, whereas the Larch (*Aphidecta obliterata*), Eyed and Striped (*Myzia oblongoguttata*) Ladybirds prefer conifers. Lower down, the 24-spot (*Subcoccinella vigintiquattuorpunctata*) and 16-spot (*Tytthaspis sedecimpunctata*) Ladybirds prefer open grasslands and rough areas of longer grass.

Other ladybird species are far more local in distribution, and are restricted to tiny environmental niches where the combination of habitat and weather is just right. The Five-spot Ladybird (*Coccinella quinquepunctata*),

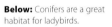

Below: Conifers are a great habitat for ladybirds.

for instance, is found only on shingle banks of the rivers Wye, Spey and Ystwyth. A few species reach the northern edge of their European range in the British Isles: the Adonis (*Hippodamia variegata*) and 11-spot (*Coccinella undecimpunctata*) Ladybirds like it warm and peter out as you go further north.

Worldwide, the same basic pattern holds true. A handful of species (including our own native Seven-spot Ladybird and the Harlequin) have cosmopolitan distributions, having spread naturally or by people either accidentally or on purpose. At the other end of the scale, many species – especially the tiny inconspicuous ladybirds – are thought to be restricted to single countries, or parts of countries. Unsurprisingly, given their long isolation from other landmasses, Australia and New Zealand are home to many of these. The Yellow-haired (*Adoxellus flavihirtus*), Large Leaf-eating (*Henosepilachna guttatopustulata*) and Minute Two-spotted (*Diomus notescens*) Ladybirds, for example, are all restricted to Australasia.

Although Coccinellidae may not be the most speciose family, the diversity (of size, habitat preferences, dietary preferences and colouration) and adaptability of ladybirds make them one of the most successful beetle groups worldwide. Wherever you are in the world – country or habitat – ladybirds won't be far away.

Below: Ladybirds are often moved around by people, either accidentally or on purpose.

Ladybirds Across Time and Space

The exhibitionists of the insect world, brightly coloured ladybirds are some of our most recognisable invertebrates today. Finds at archaeological sites dating back to the last ice age suggest that our ancient ancestors would have been just as familiar with the beetles, whose evolutionary roots can be traced back beyond the dinosaurs. Beneath their finery, ladybirds are beetles, and they have been a very long time in the making. With millennia of evolutionary history behind them, they remain a hugely diverse and impressively successful group.

Insect evolution

Ladybirds – in fact, insects in general – do not tend to fossilise well due to their lack of hard, long-lived bones. A thin chitinous exoskeleton just doesn't cut it! Most beetle remains from anything but the very recent past are limited to those preserved in amber and what palaeontologists call Lagerstätten sites, which are sedimentary rocks formed from low-oxygen mud, such as that typically found around the edges of ponds and lakes. This anoxic environment is so harsh that even bacteria struggle to survive, delaying decomposition long enough that any dead animals and plants falling into the mud are preserved exceptionally well as it solidifies.

Despite the patchy nature of the fossil record, we do have a reasonable idea of how ladybirds evolved. The earliest common ancestor of insects and other arthropods was probably a segmented worm-like animal with antennae and small limbs, which made a living approximately 540 million years ago by eating its way through marine sediments. These creatures began to

Opposite: Trilobites are superficially similar to ladybirds; however, ladybirds' different anatomy makes them much less likely to fossilise.

Below: It is believed that a segmented worm-like animal was the earliest common ancestor of insects.

Right: Early arthropods, such as trilobites, eventually gave rise to modern crustaceans and insects.

evolve external skeletons as a defence against marine predators, which were becoming faster and deadlier as the concentration of oxygen in the atmosphere increased.

In parallel with the development of this armour plating, jointed legs and appendages came into being, to make the bulky exoskeletons more manoeuvrable and to increase dexterity for catching and handling prey. Insects are thought to have evolved from early crustaceans, so fossils of these animals – which include the trilobites – give us a general guide to the kind of adaptations required to survive in the hostile prehistoric seas. The 508 million-year-old Burgess Shale deposits in British Columbia, Canada, contain a weird and wonderful range of fossils, including arthropods that may be some of the ladybird's earliest relatives.

Around 100 million years later, these ancestral arthropods began to leave the sea. The tracks some of these creatures left on beaches – and occasionally even the fossilised animals themselves – can still be found today, providing a tantalising glimpse into the past. To survive on land, the arthropods began to develop systems to breathe air directly – including tubes known as trachea to transport air through the body – rather than absorbing it from the water. This allowed the proto-insects to leave the increasingly crowded and dangerous sea to forage along the tidelines and beaches, before they gradually moved further inland.

Another leap forward – 80 million years or so, to 360–350 million years ago – saw springtails, bristletails and silverfish rampant among the undergrowth, breaking down plant material to produce soil. Around this time, the first wings evolved – on the back of an ancient insect. The scarcity of fossil arthropods means we can't be sure which species was the first to fly, but the earliest known example of a winged insect is the trace fossil of a mayfly landing site dated back 310 million years. It's clear from the rock that this insect touched down on the surface of what was then mud, before flying away – there are no footprints on either side of the imprint. Sole owners of the skies for 90–150 million years, the insects could – and did – colonise the prehistoric forests, from the treetops to below ground level, forming a huge range of diverse new species and groups.

The oldest beetle-like fossils date from the Lower Permian, around 280 million years ago. The key adaptation of beetles – hard, armoured elytra to protect the delicate flight wings – first turns up in the fossil record around 20 million years later. Then, around 252 million years ago, beetles got the opportunity they had (unknowingly) been waiting for. The Great Dying, or end-Permian mass extinction event, which wiped out 90–96 per cent of the species alive at the time, seems – somewhat counter-intuitively – to have given beetles their big break. Surviving the disaster, they were poised to take over. The first known fossilised beetles date to around 240 million years ago, during the Triassic period. The diversity of forms already present at this point suggests beetles were undergoing an adaptive radiation, or large series of speciation events, as they evolved to fill the environmental niches that had suddenly, cataclysmically become available.

Enter the ladybirds

Beetles continued to evolve alongside the dinosaurs, with a significant increase in species and family diversity during the Jurassic period, 200–145 million years ago. The evolution of flowering plants around 125 million years ago saw a further diversification as more specialised beetles evolved. Herbivorous beetles began to split off from each other as they adapted to feed on different

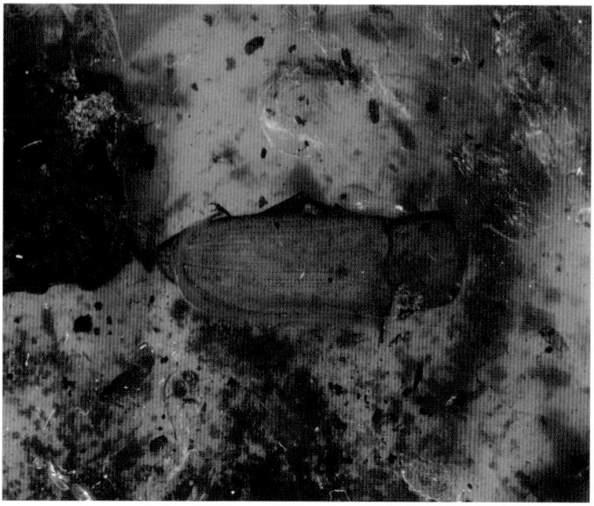

Right: Early beetles have been found in amber and fossilised. It can be challenging to identify a species from the remains, as most fossils are different body parts.

newly evolved plant species, and on parts of plants that hadn't existed before, such as pollen and seedpods. Carnivorous species, detritus-feeders and fungivores began to increase in number as well, specialising in a similar fashion.

Most beetle fossils are not of whole individuals but of disarticulated body parts such as heads or elytra, making identification to species level often very difficult. Most early ladybirds seem to have been more similar to the modern-day inconspicuous ladybirds in general appearance – in other words, mostly small and brown or black. Because of this, there are no definitive ladybird fossils from the Jurassic period, only disputed signs of potential early Coccinellidae. The earliest confirmed ladybirds – found preserved in amber – date from the Lower Eocene, 56–48 million years ago. Many of these are similar enough to current species that they can be slotted neatly into modern-day genera, suggesting that the family evolved some time previously without traces being preserved. However, ladybirds don't seem to change much throughout time: fossils in Quaternary deposits dating back 2.4 million years are often virtually identical to living ladybirds, and in many cases can be easily identified as species that still exist today.

Arrival in Britain

There are currently around 47 species of ladybird in the UK, at least five of which have become incontrovertibly established here only during the past couple of decades. Most of the remainder colonised Britain as the ice sheets retreated for the last time around 11,000–10,000 years ago, before the flooding of Doggerland and the land bridges connecting Britain to the Continent around 8,000 years ago. For most of these species, Britain was not terra incognita. The northwards and southwards movements of the polar ice sheets over thousands of years had periodically revealed the British landscape for colonisation, before burying it again. Ladybirds – along with a range of other species, including primitive humans – moved in when conditions were good, then retreated southwards as the glaciers advanced again, often repeatedly over thousands of years.

Digging through the silty soils beneath our feet has revealed intriguing glimpses of the ladybird fauna of this lost Britain. A single silt deposit in southern England, dated back 43,000 years, has revealed 248 beetles, including the Seven-spot, 11-spot and Water (*Anisostica novemdecimpunctata*) Ladybirds, along with *Scymnus frontalis*, all four of which still occur here today. They were not alone in ancient Britain, however, and species found in other deposits provide evidence of the changing conditions. Sediments laid down around 40,000 years ago by what is now the river Thames entombed the ladybird *Ceratomegilla ulkei*. Today, this species is found only in north-western Canada and at high altitude in eastern Asia. This indicates that the climate in Britain at the time the ladybird lived here was colder than now, equivalent to that of modern-day northern Scandinavia. This is backed up by UK finds of other subarctic ladybird species, such as *Anisosticta strigata* and *Hippodamia arctica*, and the fact that all the subfossil remains of the 11-spot Ladybird (found here in sediments laid down 40,000–25,000 years ago) are the form *confluens*, now common only at the northern edge of the species' range.

Below: *Scymnus frontalis* is one of the oldest ladybird inhabitants in Britain.

Right: Even today, ladybirds must be able to withstand very low temperatures to survive the cold winters.

The commonest of the early British ladybirds seems to have been *Hippodamia arctica*. A 4–5mm-long (3/16in) black-and-yellow species with linear squiggle markings rather than spots, it is now found only in arctic and subarctic tundra. In Canada, it occurs in the far north of Quebec and Labrador, as well as in British Columbia, the Yukon and Northwest Territories. In Europe, it is found in the far northern reaches of Scandinavia and the Russian Arctic coast. In prehistoric Britain, it seems to have favoured cold conditions around the southern edges of the ice sheets. The fossil record indicates that the species was apparently widespread across Britain from 40,000–25,000 years ago, before vanishing here as the temperature fell. It returned around 13,000 years ago and spent several hundred years moving northwards through Britain, following the glaciers as they retreated. A couple of thousand years later, during the Younger Dryas period (12,900–11,700 years ago), temperatures fell and it headed south again, before moving north and then vanishing for good as the climate warmed.

Although the increasing temperatures ousted *Hippodamia arctica* from our shores, they allowed most of our modern-day ladybirds to colonise Britain from the south. Some of the earliest dated remains of these incoming species have been retrieved from the Sweet Track in Somerset, the second-oldest timber causeway known in Britain, built almost 6,000 years ago in 3807 BC. The track runs through acidic peat, which preserved both the structure itself and organisms along and in it – including *Scymnus* and *Chilocorus* ladybird species.

Below: This is a reconstruction of the Sweet Track, a timber causeway used by ladybirds almost six million years ago.

British ladybirds today

The modern-day British Isles has around 47 resident species of ladybird. These are sorted into four subfamilies: Coccinellinae, comprising the big, brightly coloured aphid-eaters; Chilocorinae, including the red-spotted black species that eat scale insects; and Epilachninae and Coccidulinae (sometimes divided into smaller subfamilies such as Scymninae), comprising the generally smaller, hairy species. All members of the subfamily Coccidulinae, plus one species (*Platynaspis luteorubra*) in Epilachninae, are inconspicuous ladybirds, while all of the species in Coccinellinae and Chilocorinae, plus two species – the Bryony (*Henosepilachna argus*) and 24-spot Ladybird – in the family Epilachninae, are true ladybirds. Worldwide, the distinction between true and inconspicuous ladybirds tends to be dropped, and all are simply referred to as ladybirds (or local equivalents such as ladybug).

The number of species found in Britain does fluctuate, both up and down. *Nephus bisignatus* is now thought to be extinct in the UK, and there are only old records of *Exochomus nigromaculatus* and the 12-spot Ladybird

Below: The 13-spot Ladybird has a chequered history of extinction and recolonisation in Britain.

(*Vibidia duodecimguttata*). The 13-spot Ladybird (*Hippodamia tredecimpunctata*) went extinct as a breeding species in the twentieth century, and there were only sporadic records of immigrant individuals until larvae were found in Devon in 2011 and Sussex in 2017. Time will tell if the species is back for good.

In general though, the second half of the twentieth century was pretty positive for ladybirds, and several species expanded their ranges. The Orange Ladybird (*Halyzia sedecimguttata*) went from being a species largely restricted to ancient oak woodlands, to one of the most widespread and abundant nationwide after it discovered it could eat the mildew on Sycamore (*Acer pseudoplatanus*) as well as that growing on oak leaves. *Nephus quadrimaculatus*, once known in the UK almost exclusively from Suffolk, started to be found across the south-east and then further afield. Additionally, new or expanded populations of the Scarce Seven-spot (*Coccinella magnifica*) and Five-spot Ladybirds were discovered.

Several new ladybird species have also colonised Britain over the past century. The first to arrive was the Cream-streaked Ladybird (*Harmonia quadripunctata*), which turned up in west Suffolk in 1937 and has since spread slowly across the country, colonising pine trees and other conifers. This was the only arrival to establish itself for half

Below: The Orange Ladybird is now one of our most common species after discovering that it could eat mildew on Sycamore as well as mildew growing on oak leaves.

Left: The Bryony Ladybird has gradually colonised Britain since establishing itself in Surrey in the late twentieth century.

a century, before a series of others began to arrive in the closing years of the twentieth century.

The tiny but attractive *Scymnus interruptus* was first found in Britain in 1986, but there was then a 10-year gap before a second individual was discovered. There is a series of records of the species after 1996 from the south-east, mainly in coastal areas, suggesting it was occurring as an immigrant rather than breeding here. That changed

Below: The first British Cream-streaked Ladybird was found in Suffolk shortly before the Second World War.

in 2011–12, however, when both adults and larvae of the species were found across several sites in Berkshire and south Oxfordshire, and it has since been found widely across southern England in grassland and scrub, as well as in gardens.

The small, inconspicuous brown ladybird *Rhyzobius chrysomeloides* was found on a Surrey motorway bank in 1996, and the red and gunmetal-grey *R. lophanthae* was first found in the wild in Britain in the same county in April 1999. Both expanded their ranges quickly and are now widespread in conifers (mainly pines and cypresses, respectively) across southern England. Although there are few Welsh or Scottish records to date, it is likely only a matter of time before the ladybirds turn up. Very recently, the two *Rhyzobius* species have been joined by a third, the all-black *R. forestieri*, although this currently seems to be restricted to central London. It is striking that so many of the colonising ladybird species seem to prefer conifers. Indeed, it may be the widespread planting of these trees outside their natural range in the UK, both ornamentally and in plantations, that has made the ladybirds' colonisation possible.

The much larger Bryony Ladybird turned up in Surrey in the mid-1990s. Unlike the other ladybird immigrants, it has spread slowly and is still mostly found in the county – although a population became established in Oxfordshire in 2010, and another in Essex more recently. Both populations appear to be replicating the growth of the Surrey population: locally abundant, but spreading slowly.

The most famous ladybird invader is, of course, the Harlequin. Originally native to temperate Asia, this species was introduced to Continental Europe and has since spread to the UK, arriving and establishing itself here in 2003–04. The species has since rampaged across the country and is now the most abundant ladybird species in most of England and Wales, although it appears to be less common in Scotland. It has largely replaced the Two-spot and surpassed the Seven-spot as the commonest UK ladybird species; the full story of the devastation it has wrought in doing so is told on pages 86–89.

Ladybirds worldwide

The sheer diversity of ladybirds has allowed them to occupy virtually every terrestrial habitat, and they are abundant all over the world. Several species – such as the Seven-spot and Three-banded (*Coccinella trifasciata perplexa*) Ladybirds – can be found on the tundra in the high Arctic regions of Russia and Alaska. In contrast, others – such as the Twice-stabbed Cactus Ladybird Beetle (*Chilocorus cacti*) – are entirely at home on the sparse vegetation of arid regions across Asia, Africa, the Americas and the Middle East.

Several species overwinter at or near the snow line on mountain peaks. Indeed, the Pink Spotted Lady Beetle (*Coleomegilla maculata*) is so well adapted to spending the winter in these conditions that overwintering adults survive better when covered by snow. By contrast, the tropics are some of the richest regions for ladybirds. The Indian subcontinent has around 520 known species, and in the US state of Hawaii alone, the number of ladybird species successfully introduced (46) is virtually equivalent to the total number of species in the whole of the UK.

Many ladybird species are restricted in distribution, but others can be found widely across regions or continents. Probably the two most widely distributed species globally are the Seven-spot and Harlequin Ladybirds, found in at least 73 and 57 countries, respectively. The ranges of both of these species have been artificially increased, as they have been introduced to new countries as a form of biological control for pest insects. Several other species have also been moved around the globe by people for this reason – Australian ladybirds such as the Vedalia Beetle (*Rodolia cardinalis*), for instance, can be found across Europe and North America (see pages 56–59).

Above: Pink Spotted Lady Beetles preferentially overwinter on mountains.

Above: Several ladybird species can be found even in very arid desert conditions.

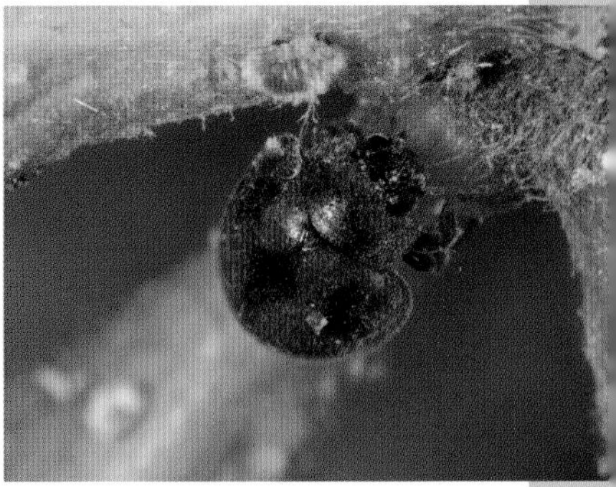

Above: The Vedalia Beetle has been shipped all over the world as a biological control agent.

Appearance isn't everything

Not everything that looks like a ladybird is actually a ladybird, and not all Coccinellidae look like ladybirds! The leaf beetles in particular can look very similar at first glance. Several species of *Cryptocephalus* leaf beetles are brightly coloured and some are spotty, although all are quite elongated and have grooves in their elytra. The Poplar Leaf Beetle (*Chrysomela populi*) can also easily be mistaken for a spotless ladybird: it has bright red elytra, and a jet-black head, thorax and underside. However, at 10–12mm (3/8–1/2in) in length, it is larger than any British ladybird species and has longer antennae.

Tortoise beetles in the genus *Cassida* don't mimic ladybird colouration (they're generally green and brown) but they do match the shape of a ladybird fairly well, aside from being much flatter. Essentially, they look like a ladybird that someone has sat on! The undisputed champion of ladybird mimicry, however, is the so-called False Ladybird (*Endomychus coccineus*).

The inconspicuous ladybirds are small, hairy, and black or dark brown with indistinct, dark red splodges. Nonetheless, they can be picked out without too much difficulty after a bit of practice – their general shape and the way they move meshes with finer details such as the shape of the antennae and palps into a signal that says 'ladybird'.

Until recently, no discussion of unusual ladybirds would have been complete without mention of the bizarre *Cleidostethus meliponae*. This species has only ever been found in East Africa in nests of the stingless bee *Melipona alinderi*. Both blind and wingless, the beetle appears to be highly adapted to life in the bee nests but, unfortunately for our purposes, a taxonomic decision in 2001 removed it from the Coccinellidae and moved it to a sister family, the Corylophidae.

Below: Despite looking like a slightly squashed ladybird, this is actually a tortoise beetle from the Genus *Cassida*.

Ladybird lookalikes

The prize for mimicry of ladybirds, at least in the UK, must go to *Endomychus coccineus*, in the so-called handsome fungus beetle family (Endomychidae). Found across Britain and Europe, it looks so similar – smooth, rounded and bright red, with four big black spots – that it rejoices in the common name False Ladybird.

False Ladybirds live beneath bark and on tree stumps, where they feed on fungi, so are unlikely to be found in the same habitats as ladybirds, and no British species has four black spots (at least normally). This makes them difficult to mistake for real ladybirds in some ways, but in others the mimicry is surprisingly effective. The False Ladybird even produces fluid from joints in its legs when disturbed, as do ladybirds (see page 73), although this is pink rather than the typical yellow fluid produced by ladybirds, and it is not known whether it is distasteful or toxic, as in ladybirds, or merely mimicry.

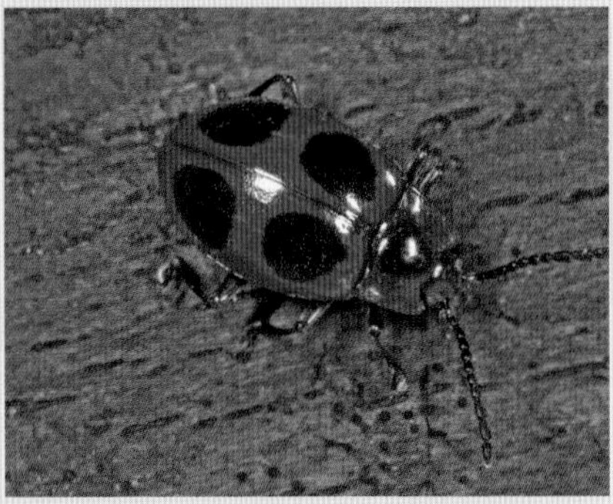

Above: Often mistaken for a four-spotted ladybird, this is actually a False Ladybird from the Endomychidae beetle family.

Left: Inconspicuous ladybirds, such as this *Scymnus interruptus*, look less like a stereotypical ladybird than some of the mimic species.

The Ladybird Year

A ladybird's life cycle starts as one of a batch of oval yellow eggs, laid sometime in the summer. Hatching into a tiny black larva, the ladybird eats and eats and eats for a month before pupating. About 10 days later, the front of the pupa splits open, and the new adult ladybird appears for the first time. It, too, spends most of its time eating, fattening itself up for the winter on pollen, nectar and fruit, as well as aphids and other pest insects. After sleeping through the winter, the ladybird disperses to find food and a mate, and the whole cycle begins again.

Emergence and dispersal

Ladybirds are not among the earliest risers of the insect world. As predators, they are better served by sleeping in and allowing their prey insects to build up their populations before they, in turn, begin to stir. Once the initial few sunny days of spring arrive in the UK, in March or early April, the first ladybirds will begin to appear too. Usually, the first to be seen outside are the Pine (*Exochomus quadripustulatus*) and Kidney-spot Ladybirds, sunning themselves on tree trunks. Mainly black in colour, they absorb more of the sun's energy than do lighter species, so can be active at lower temperatures.

It's not long before the rest of the ladybird species appear. The sun-dappled trunks of deciduous trees remain a hotspot until the emerging leaves shade them out, and early growth in nettle patches is often a prime basking location for Seven-spot and 14-spot Ladybirds. While nights remain cold, some species – including the Pine and 11-spot Ladybirds – will deliberately seek out sunny spots in the morning, climbing vegetation to sit at the top, where they can warm up despite being exposed.

Fairly soon, a ladybird's thoughts turn to food. There is generally none to be found at the overwintering site, so

Opposite: The life cycle of a ladybird begins with a larva hatching from an egg.

Below: Many ladybird species overwinter in huge aggregations.

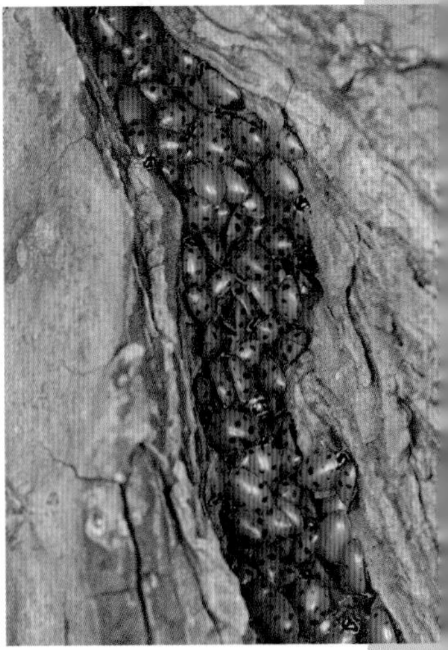

Above: Pine Ladybirds often overwinter in exposed locations, such as tree trunks.

Below: Seven-spot Ladybirds overwinter in leaf litter, but climb up plants to bask in the spring sunshine.

spring is a time of dispersal, with the beetles undertaking short-, medium- and long-range flights to seek out colonies of prey insects. They are clumsy but efficient flyers. Their elytra open outwards and forwards, and are held in place while the membranous hindwings are unfurled. With the wings fully extended, there is a brief pause – as if the ladybird is double-checking everything is ready – before, with a little jump, it launches itself skywards. Or at least generally skywards: if something unexpected happens part

Above: When preparing to fly, flight wings unfurl from beneath the raised elytra. These hindwings then flap rapidly to propel the ladybird through the sky.

of the way through this process, such as the launch site moving, the ladybird will usually fall, wings splayed, rather than aborting take-off.

If the take-off goes well, the hindwings will flap at around 85 beats per second, propelling the ladybird through the sky. It is not clear what role the elytra play in flight, but they are clearly of some use. Experiments have shown that a ladybird without elytra cannot fly, even when the main wings are untouched. Radar has detected Harlequin and Seven-spot Ladybirds more than 1km (3,300ft) up in the sky and moving at more than 60kph (37mph), so long-distance movements of 100km (60 miles) or more are feasible, especially when ladybirds ascend into fast-moving air masses such as storms.

Mating

Above: This Pine Ladybird may be attracted to the much larger Harlequin, however, he is definitely wasting his time.

Below: Ladybirds can mate many times, and often for long periods.

Aside from food, dispersing ladybirds are on the lookout for mating opportunities. The beetles don't display any elaborate mating or courtship behaviours: males appear to find females more or less by walking into them.

A male ladybird's response to bumping into anything vaguely ladybird-shaped and about the right size is to climb on top of it. Female ladybirds produce a cocktail of chemicals from pores in their elytra that specify their species and sex. Once on top of a potential female ladybird, the male stops to check for these chemicals. If the chemicals are wrong (because the other ladybird is of a different species or is another male), or if there are no chemicals (because the ladybird is dead or actually another object altogether), then the male breaks off his courtship attempt and continues on his way.

If the male has successfully found a female of his own species, however, he will turn himself around until he's facing the same way as she is, then he lines himself up for mating. The female's usual response is to try to repel him, withdrawing her abdomen and trying to shake him off to gauge whether he is a worthy mate. If he persists and she finds him suitable, she will allow him to mate with her, which can take as little as 15 minutes or as long as nine hours. Female Two-spot Ladybirds will mate every day during the breeding season when the weather is right, or around 30 times in total, on average.

Eggs in abundance

A day or so after they first mate, female ladybirds begin to lay eggs. Adults and larvae eat the same food, so a well-fed adult female that has just mated should be in the right place to lay her eggs where they have a good chance of survival. However, eggs and young larvae are vulnerable to other ladybirds, both through direct predation and through competition for food. Ladybirds all produce recognition chemicals from their exoskeletons, including their legs, leaving a scent trail behind them. Like bloodhounds tracking criminals, ladybirds can detect where others have walked, and they will actively avoid areas with excessive ladybird traffic when looking for egg-laying sites, even if there are plenty of aphids around – the risk to their offspring is just too great.

Below: Ladybirds will lay batches of up to 50 yellow rugby-ball-shaped eggs at a time. They often leave them glued to the surface of a leaf.

This site selection is as far as parental care goes in most ladybirds. Once the female has found a site with aphids or other food, but little sign of other ladybirds, she will begin to lay her eggs. These are the shape of a stretched rugby ball, in varying shades of yellow, and laid end-on in clutches of usually 10–50 eggs. The female will repeat this as frequently as she can for the rest of her life, and may produce an astounding number of offspring: one female Harlequin Ladybird laid 2,200 eggs in captivity.

The tropical Asian Giant Ladybird (*Megalocaria dilatata*) takes things a step further. As a larva, it frequently feeds on aphid colonies defended by ants, so the adult female doesn't just pick a site for egg-laying (the ends of bamboo shoots), but also extrudes a series of thick, sticky rings of gluey gloop around the shoot, preventing the ants reaching the eggs. The hatching ladybird larvae seem able to dissolve the glue without getting stuck.

Larval ladybirds

Ladybird eggs gradually darken until, just before hatching, they are almost black, with the new larva visible through the shell of the egg. At this point, around five days after laying and just before hatching, it becomes clear that a few of the eggs in some batches are not going to hatch: they remain bright yellow, with no sign of a developing, darkening larva inside. These dead eggs may well be the result of inbreeding, or of diseases passed down from the mother, but it is also thought that in some cases ladybirds may deliberately lay unfertilised eggs. These eggs, known as 'nurse' or 'trophic' eggs, are there solely to provide food to newly hatched larvae, which eat their own eggshells and any unhatched eggs as they hatch en masse from each clutch.

Larvae hatch from their eggs by bursting through the eggshell using hatching spines on their heads, either splitting the egg down the middle or at one end. Once free of the egg, the larvae rest for an hour or so while their exoskeleton hardens, and then are free to move around, eating anything they can.

Many larvae don't survive this first test and die without ever successfully eating. The survivors eat and eat until they come up against the major problem of exoskeletons: they don't stretch. To continue to grow, a larva has to change its skin.

Below: Ladybird eggs will darken shortly before hatching. Any remaining bright yellow eggs have not been fertilised and will be eaten by the newly hatched larvae.

Left: The shed larval skin can still be seen attached to the back end of this ladybird pupa.

The larva does this by first growing a more capacious skin beneath its existing exoskeleton, soft and wrinkled to allow it to fit. The larva then stops feeding and attaches itself to a leaf, using a pad at the end of the abdomen to hold on. The old skin splits open around the head and peels back from the top of the thorax and abdomen, allowing the larva to walk forwards out of the old, shed skin. Even the linings of the foregut and spiracles are replaced, as they are attached to the outer skin. The larva then waits for its new skin to harden, during which it darkens from fluorescent orange to more normal ladybird colours, before resuming feeding. This process is repeated three times as the larva grows to full size. The stages between skin-sheds are known as instars, and virtually all ladybird larvae have four instars between hatching and pupation.

Ladybird larvae look nothing like adults, instead resembling tiny six-legged crocodiles that scuttle at speed across leaves and tree trunks. Most are simply black or dark grey at first, gradually developing colours and markings as they grow. First- and second-instar larvae are very difficult to tell apart, but third- and fourth-instar larvae have usually developed enough spines, hairs, warts and coloured patches to be identifiable to species level. As in caterpillars and butterflies, the colours and patterns of larval and adult ladybirds are completely different. In the UK, only one species, the Orange Ladybird, has the same colour scheme of orange, yellow and white across egg, larva, pupa and adult.

Below: This half-grown larva is an Orange Ladybird; it is one of the few species to retain the same colour scheme throughout all stages of the life cycle.

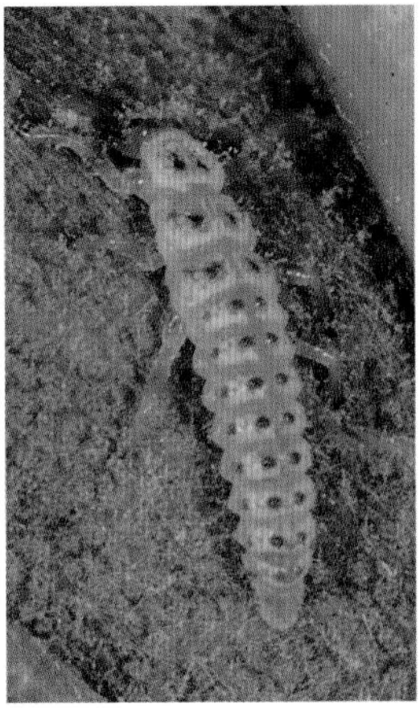

Pupation

The fully grown final-instar larva has one last job to carry out: pupation. As when shedding its skin, the larva stops feeding and grips a leaf tightly with the pad at the end of its abdomen. It arches its body into a hunched posture and darkens in colour. The larva remains like this – in what is known as the prepupal stage – for several hours, before wriggling out of its larval skin for the final time, revealing the pupa beneath. On close examination, the pupa can be seen to have all the features of the adult ladybird but with the proportions slightly – or majorly – off. There are six legs folded up underneath, with tiny wing-buds on top and antennae pressed against the sides. The pupa is initially a bright, fluorescent orange-yellow, but it soon darkens to its normal colour. Depending on the species, this may be black, orange, white or another colour, but there is usually no colour coordination with either the larva or adult.

Right: During pupation, the last larval skin splits open and the pupa wiggles free.

Left: After emerging from the larval skin, the pupa quickly darkens from fluorescent yellow to its normal colours.

Most species exit their larval skins almost completely, just leaving the remains of the shed skin wrapped around the back end of the pupa, where it is attached to the leaf. However, species in the subfamily Chilocorinae, such as the Pine and Kidney-spot Ladybirds, merely crack open the final larval skin down the middle of the back, splaying the two sides open to reveal the pupa inside, and allowing the impressive spines of the larva to continue their protective work. Some species of *Scymnus* and *Hyperaspis* take things a step further. Although the final larval skin is completely disconnected from the pupal skin inside, the outside is not broken until the adult ladybird forces its way out.

Below: In some species, such as these Kidney-spot Ladybirds, the larval skin splits open down the back to reveal the pupa within.

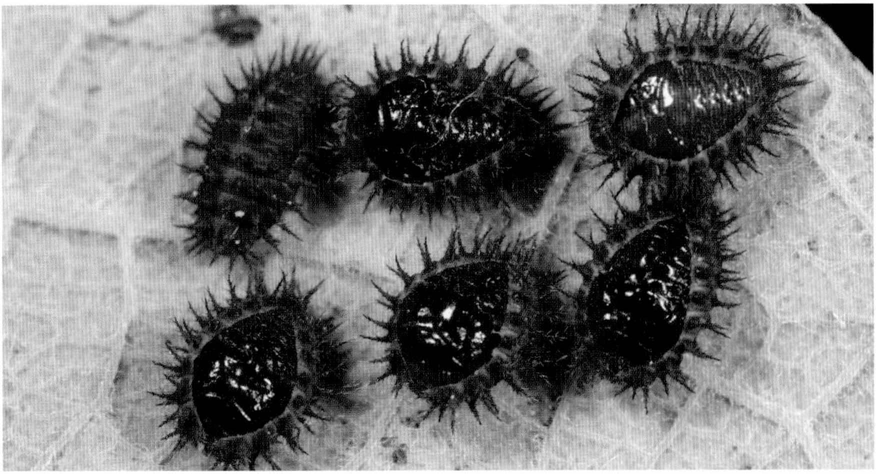

Adult emergence

About a week or 10 days after pupating, the ladybird is ready to emerge as an adult at last. The skin of the pupa splits open across the top of the thorax (the front of the pupa, because it's so hunched over), and the newly adult ladybird simply walks forward out of the pupal case. It rests, pumping haemolymph (insect 'blood') into the wings and elytra to expand them, while the rest of the exoskeleton hardens and darkens. Young adults look fresher and brighter than their parent's generation, with colours gradually darkening over the lifespan of the adult. Adult ladybirds do not grow, however – all the growth takes place at the larval stage.

Right: This newly emerged Orange Ladybird will gradually darken as it matures.

Right: The legs, wing-buds and abdominal segments of the eventual adult ladybird can clearly be seen on this pupa.

In the UK, only a handful of species will go through this process more than once a year, although multiple broods are more common in warmer countries. For most British species, emergence from the pupa fires the starting gun on a race to consume as many calories as possible before overwintering.

In 'ladybird summers' such as that of 1976 (see box, page 67), the mass emergence of hungry new adult ladybirds coincides with the harvesting of field crops in early August. This combination of factors wipes out the few remaining aphids in fields, so the ladybirds take to the skies in search of more. In coastal areas – particularly in East Anglia, which has a high proportion of crop fields, and coastlines to the north, east and some of the south – these dispersing ladybirds reach the sea and can go no further, building up to vast numbers in seaside resorts.

Desperate for food, most ladybird species broaden their diet at this time of year. Some may go as far as attempting to bite humans – although it would take an unusually powerful bite to break the skin. Others begin to feed on berries or nectar. In North America, the introduced Harlequin Ladybird has become a pest in vineyards thanks to its habit of feeding on grapes and then hiding in the centre of the bunch. When the grapes are picked and then trampled, the ladybirds are crushed as well, and their defensive chemicals taint the wine that is produced. A single Harlequin is reputed to make up to 5 litres (9 pints) of grape juice unusable.

Below: Sugar-rich fruit juice attracts ladybirds in the early autumn, sometimes with unfortunate consequences; when the grapes are picked and trampled, the ladybirds can be crushed as well.

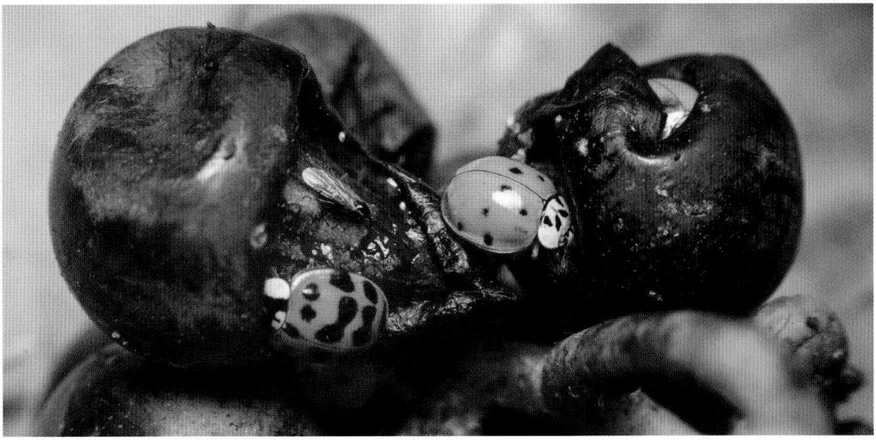

Overwintering and dormancy

Ladybirds overwinter as adults, so their new year is marked by several months of uninterrupted sleep rather than being heralded in by fireworks and the chimes of Big Ben. Technically, they don't hibernate (the scientific definition of hibernation involves controlling body temperature in a way that insects are not capable of), but instead go through a period of winter dormancy. The beetles are usually triggered to find an overwintering site by a combination of day length, temperature and food availability. In some species, just one factor is important; for instance, the Two-spot Ladybird seeks out overwintering sites once the day length drops below 14 hours. Most species are set off by two or more factors working together, such as a certain number of days below a particular temperature and a reduction in day length beyond a certain number of hours.

This autumn dispersal phase sees different species seeking out different overwintering sites. Many ladybirds, including the Cream-streaked and 18-spot (*Myrrha octodecimguttata*), overwinter on their host trees. Others, such as the Orange and Seven-spot Ladybirds, move short distances to overwinter on evergreen plants or in leaf litter. Several species move into conifers to spend the winter

Below: Ladybirds often overwinter between the needles of conifers.

Overwintering aggregations

By far the most spectacular overwintering behaviour is seen in species such as the Harlequin and the North American Convergent (*Hippodamia convergens*) Ladybirds. These species overwinter in huge aggregations, sometimes numbering in the millions per site, with individuals travelling long distances to reach these communal areas.

Taking to the air, the ladybirds first home in on visually distinct areas such as hills or pale buildings, especially on the edge of wild areas. Large numbers can accumulate here – on the tops of hills, they can feel like a downpour of ladybirds! They are then guided by smell as they follow the scent trail of other ladybirds, either those laid recently or old trails from previous years. They also show an innate tendency to join existing groups of ladybirds, and a general desire to tuck themselves into corners and crevices. This is common across ladybird species – bamboo canes, for instance, can be full of a wide range of overwintering ladybirds between October and April in the UK.

In the natural order of things, Harlequin Ladybirds overwinter in groups in caves and under rocks on hilltops and mountains, as do many other species around the world. Increasingly, however, these alien introductions can be found overwintering in artificial caves: houses. Groups numbering from a handful of individuals to thousands accumulate in window frames, on curtains or in the corners of rooms, squeezing through tiny holes such as keyholes and air bricks.

Although they are not pests at the level of bedbugs or fleas, these overwintering groups of Harlequins are often regarded as a nuisance. They will remain active in the warmth of a centrally heated home, and spend all of winter flying into lights or dropping into things. When disturbed, they can also produce defensive chemicals that stain soft furnishings yellow, and produce a chemical smell that many people find unpleasant.

Above: Overwintering congregations can number in their millions. Here, Convergent Ladybirds have engulfed a tree stump.

tucked in tight at the base of the needles – Christmas trees brought indoors in December are the source of ladybirds found flying around living-room lampshades or battering themselves against windows.

Once they are tucked up for the winter, most ladybirds will remain immobile until spring comes, heralded by increases in temperature and day length. They can live off their stored fat for a surprisingly long time – nine months or more – and are amazingly hardy. Incredibly,

Above: Harlequin Ladybirds often overwinter indoors, in large groups tucked into corners.

Hippodamia arctica ladybirds have been recorded frozen solid in ice at -20°C (-4°F) for three months, before being resurrected: once defrosted, the beetles became active inside six hours, began mating in less than three days and were laying eggs three days after that.

Most ladybirds are not quite so hardy, and over winter, large numbers are lost to predators, disease, cold weather and many other causes. However, enough find secure, sensible hiding places that spring is still heralded by the bright flash of elytra in the dead, brown detritus left by winter.

In warmer countries, the hotter seasons can be just as hard for ladybirds. In parts of the world with hot, dry summers, for example, many ladybird species will have a summer dormancy period equivalent to hibernation: aestivation. With the sun baking plants to a crisp, populations of aphids and scale insects peak in spring and autumn and decline in midsummer, limiting the food available to ladybirds. These ladybirds then have a choice between dispersal to search for more food elsewhere (generally the situation in Britain – see box, page 67), or waiting it out in the hope that populations of prey insects will build up again.

An individual's decision on which course of action to take is likely to be guided by a range of factors, such as suitability of the environmental conditions for dispersal versus aestivation, prey availability and so on. Aestivation is generally not such a deep sleep as the winter dormancy, although it can be. In the Japanese subspecies of the Seven-spot Ladybird (*Coccinella septempunctata brucki*), for example, the ladybird's respiration drops to 20–50 per cent of the normal rate, indicating that it is truly dormant.

More normally, however, aestivating ladybirds are simply waiting out the unfavourable conditions. The Australian Striped Ladybird (*Micraspis frenata*; not to be confused with the British Striped Ladybird) has a summer dormancy period, which it spends in huge aggregations in eucalyptus-studded grasslands fringing the continent's desert interior. When the rains arrive and the grasslands come to life, the species becomes active once again, but there is another trigger they seem to be even more sensitive to: smoke. The grasslands are prone to burning – indeed, most of the plants are adapted to periodic fires and in some cases actually need them to reproduce. This makes sleeping in the grasslands somewhat hazardous, and dormant Striped Ladybirds will wake and attempt to run or fly away at the faintest whiff of smoke on the breeze.

Below: Summer fires can spread rapidly through dry Australian grasslands, incinerating ladybirds that don't wake up in time.

Feeding and Foraging

One thing almost everyone knows about ladybirds is that they eat greenfly and blackfly – aphids. As is so often the case with ladybirds, this generalisation has a grain of truth to it, but the full reality is far more complex. Many of the bigger, showier ladybirds do indeed eat aphids, some in vast quantities, but as a family, ladybirds have some of the most varied diets of any beetle group. Different species are specialised for eating aphids, scale insects, mites, beetle larvae, ants, whitefly, pollen, leaves or mildews, while others are generalists that eat virtually anything smaller than they are.

Aphids – greenfly and more

The vast majority of big, brightly coloured ladybird species worldwide feed primarily on aphids. These include generalists such as the Harlequin, Two-spot and Seven-spot Ladybirds, which eat virtually any aphid species; Cream-spot, Five-spot and Water (*Anisosticta novemdecimpunctata*) Ladybirds, which eat a range of aphid species but have a much more marked habitat preference (deciduous trees, shingle riverbanks and reedbeds, respectively); and specialists like the pine-dwelling Striped Ladybird, which can reproduce successfully only on a diet of giant conifer aphids (*Cinara* species). The tropical ladybird *Megalocaria dilatata* represents an extreme. Almost absurdly rounded, measuring 10–13mm (3/8–1/2in) long and 9–10mm (3/8in) wide, it is bright orange with 12 black spots spread across the thorax and elytra, the latter flaring out like a skirt at the bottom. Commonly known as the Asian Giant Ladybird, the species is found across Asia, where it feeds almost exclusively on woolly bamboo aphids (it has been estimated to eat 15,000 in its lifetime).

Opposite: Most ladybirds eat only aphids, although some have broader diets.

Below: Larvae are by far the most voracious life stage of ladybirds.

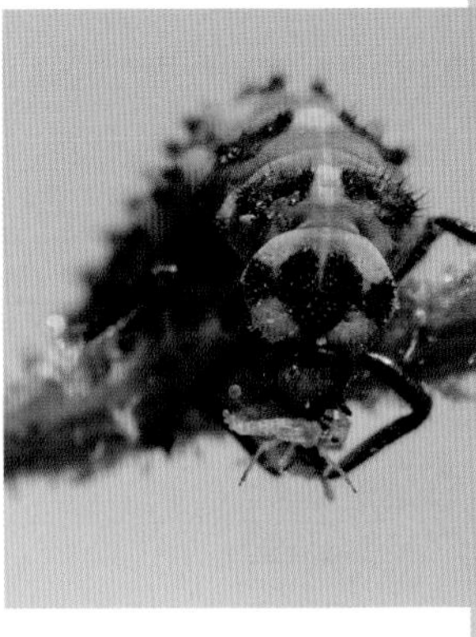

Supermarket stowaways

One of the main ways that ladybirds move long distances around the world is as accidental stowaways in fruit and vegetable shipments. Some of the earliest records of the Harlequin Ladybird in Britain and Ireland are from fruit, vegetables and cut flowers, and a variety of other species have been found in supermarkets, particularly in grapes and other fruit with a short shelf life imported from overseas.

Two species in the genus *Cheilomenes*, the Lunate Ladybird (*C. lunata*) and the Six-spotted Zigzag Ladybird (*C. sexmaculata*), are commonly found feeding on the aphids infesting citrus trees and grapevines in Asia, Africa and South America. Because of their proximity to the fruit as it is picked (especially in the case of grapes, where they sometimes rest in the middle of the bunches), both species are also found from time to time in British supermarkets. The Jamaican *Procula douie*, the South American *Eriopis connexa*, the Australian Steelblue Ladybird (*Halmus chalybeus*) and the European *Oenopia conglobata* have also all turned up in recent years. So, keep an eye out in the supermarket fruit and veg aisles!

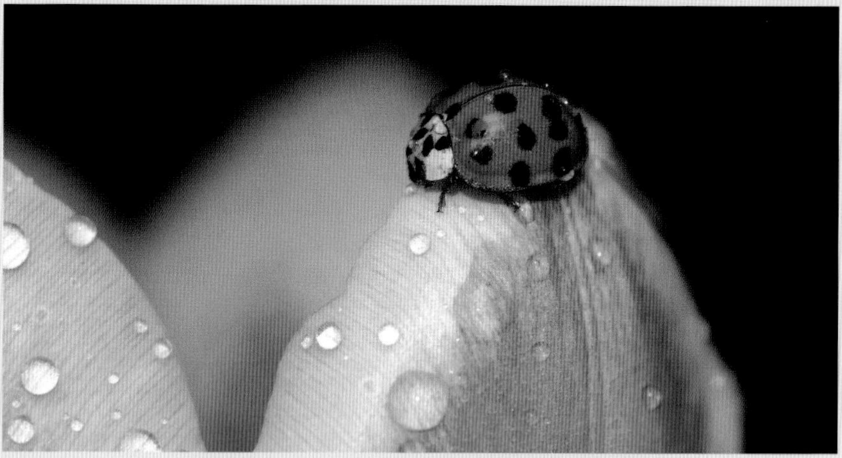

Above: Harlequin Ladybirds are frequent inadvertent travellers.

Although they might appear to be easy prey, aphids don't always make things straightforward for ladybirds. The first task for a hungry ladybird is simply to find an aphid colony. These are scattered apparently at random across the landscape, with potentially hundreds of species in an area, all with different habitat and host preferences, as well as differing palatability to different ladybird species. Ladybirds do have an ace up their sleeves though. When distressed, aphids produce a chemical known as aphid alarm pheromone, which ladybirds can detect with their

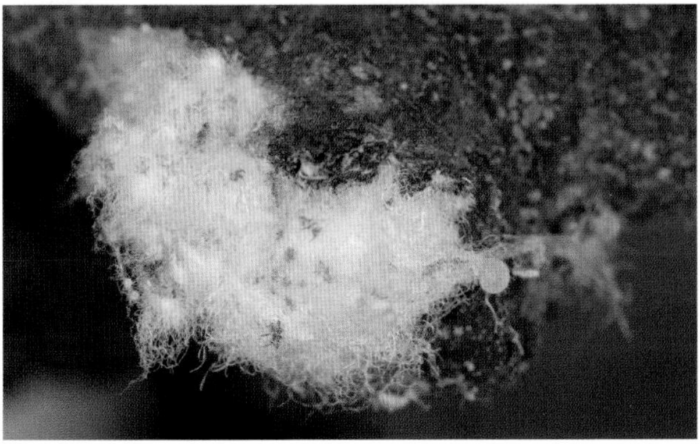

Above: Most small invertebrates are potential prey for hungry ladybirds, although aphids are a firm favourite (left). Hungry ladybirds will seek out an aphid colony that can potentially contain hundreds of individuals (right).

Left: This white fluffy material is wax. It is produced by some aphids and related insects as protection against predators such as ladybirds.

antennae and use to home in on the colony. They do this so successfully that there have even been trials to genetically engineer wheat to produce the chemical, in the hope that ladybirds will flock to the crop and hang around to eat aphids as they arrive. Unfortunately, this does not seem to have been successful.

Even once the ladybird has found an aphid colony, it's not necessarily plain sailing. Some aphids, especially the gall-living *Pseudoregma* and *Ceratovacuna* species found widely across Asia, have a specialised soldier caste.

Above: Adult ladybirds will regularly eat aphids. It is thought that a ladybird can eat up to 5,000 aphids in its year-long lifespan.

These are nymphs that, rather than growing into adults, develop a toughened exoskeleton and other adaptations, such as bigger, broader front legs. They stand guard outside the gall entrances to defend the colony, and can be quite successful at deterring attack, even killing young ladybird larvae. One Asian ladybird species, *Pseudoscymnus kurohime*, which specialises in eating these gall-forming aphids, camouflages its eggs batches from the soldiers by excreting a layer of undigested aphid over them, disguising the appearance and scent.

Additionally, not all aphids are suitable ladybird prey. Many are able to store the toxic defensive compounds produced by the plant they feed on, making them unpalatable or even poisonous to predators. For some species this is a partial effect. If it has fed on toxic Oleander (*Nerium oleander*), the Oleander Aphid (*Aphis nerii*) is rejected by the Seven-spot Ladybird, but the ladybird will happily devour the aphid if it has been feeding on Apple of Sodom (*Calotropis procera*). The Vetch Aphid (*Megoura viciae*) and Cabbage Aphid (*Brevicoryne brassicae*) are both able to collect such a quantity of toxic compounds from their foodplants that Two-spot Ladybirds have been known to die after eating just one aphid.

Above: Ladybird larvae will eat almost constantly when there is food readily available.

Other ladybirds, including the Seven-spot, are still able to eat the aphids, but neither species seems to be a pleasant meal for any ladybird.

One final hurdle for the ladybird to overcome, once it has found a colony of palatable aphids, is actually catching one. Aphids will run away, drop off leaves or fly away before a predator has the chance to attack. During an attack, aphids will kick out at predators, and frequently produce liquid wax in an attempt to gum up a predator's mouthparts. Things are even more difficult for ladybird larvae, which for much of their existence are actually smaller than adult aphids. For them, grabbing a mouthful of their aphid prey is not easy, and it is thought that a large proportion of hatchling ladybirds simply starve to death.

Eventually, the ladybird will (usually) manage to bite the aphid, jabbing its toughened mandibles through the aphid's soft exoskeleton. The ladybird will then slurp up the aphid's internals: the mandibles of many ladybird species have a groove to allow the juices to flow, and to direct them into the mouth. Small larvae will even ride around on top of the still-mobile aphid, their jaws ensuring they keep a firm grip. Larger larvae and adults go one stage further, chewing their way through until only the aphid's legs and wings are left.

Scaly situation

Ladybirds first evolved from a group of fungus-eating beetles known as the Cerylonid Series. The first proto-ladybirds were distinct from this group because they had adapted to live by predation – in particular, by eating scale insects. Adult female scale insects generally settle down on a plant and remain in one place, immobilised, as they feed on the plant's juices. With their scale-like armoured backs, they resemble an infestation of fingernails. They lay their egg masses beneath their own bodies, using themselves as a shield for their offspring, but this makes them an easy target for the handful of predators that can lever the insects up off the hostplant.

Many modern-day ladybird species – especially those in the subfamily Chilocorinae – still feed on scale insects. In Britain, examples include the Pine and Kidney-spot Ladybirds, which are black with red spots, and, like the Asian Giant Ladybird, have elytra that flare out into a lip at the bottom. The Pine Ladybird seems to be particularly fond of the Horse-chestnut Scale (*Pulvinaria regalis*), a 5mm (3/16in) brown scale insect found widely on a range of deciduous trees. The scale insect produces copious

Right: In addition to aphids, Pine Ladybrids eat Horse-chestnut Scale insects.

Above: The beautiful scale-feeding Steelblue Ladybird is native to Australia.

quantities of white waxy filaments as a protective blanket over its egg batches, in the hope of gumming up the mouthparts of predators such as ladybirds, giving the adult females something of the appearance of a miniature limpet sitting on candyfloss. This doesn't deter the Pine Ladybird, however. The female ladybirds lay their own eggs into the stringy mass, and hatching larvae eat their way through the pearly-pink scale insect eggs, emerging from the safety of the scale's protective coating as half-grown larvae festooned with waxy remnants.

Some scale-feeding ladybirds take things even further. In the USA, several ladybirds in the genus *Brachiacantha* have larvae that live underground, where they feed on scale insects in ant nests.

Perhaps the most attractive of the scale-feeding ladybirds is the Steelblue Ladybird, native to Australia and introduced to New Zealand. Adults are a beautiful, deep metallic blue all over, with tiny, pale yellow flashes at the front of the thorax and a bright red underside to the abdomen. Although each individual is only about 4mm (1/8in) long, they clump together in small groups under leaves and in nooks and crannies to sleep through the Australian winter (May–September), bedazzling their chosen foliage like glittering metallic studs.

Harnessing appetites

The desire and ability of ladybirds to eat both aphids and scale insects in prodigious numbers has made them one of the main insects of choice for the biological control of these pests in gardens, farmland and greenhouses, and even on trees planted along city streets. From at least the beginning of the nineteenth century, it was noted that ladybirds were a valuable addition to a garden pest-control strategy, and so it was not a great leap to the idea of adding more ladybirds to an aphid-plagued area – or introducing known aphid-eating species to areas without them. The first modern use of ladybirds for biological control of aphids came in 1874, when the 11-spot Ladybird was introduced to New Zealand from Britain. While the effects were not dramatic, the ladybird did successfully establish itself and is now found widely across the country. The introduction also showcased one of the pitfalls of biological control: a parasitic wasp, *Dinocampus coccinellae*, was introduced along with the ladybird and is likely to have been one of the reasons for the species' muted success as a control agent.

The first truly successful use of ladybird species for biological control was with an Australian species, the Vedalia Beetle. In the late 1880s, the California citrus

Right: The Vedalia Beetle is the saviour of Californian citrus groves, working as an effective pest-control to protect the citrus trees and save a valuable industry.

Above: Citrus orchards are particularly susceptible to pests without ladybirds present to guard them.

groves were in trouble. The Cottony Cushion Scale (*Icerya purchasi*) had recently arrived from Australia, and huge numbers of the insects were infesting orange and lemon trees across the US state, killing many and reducing the crop so much that farms began to go out of business. Desperate, the farmers spent US$1,500 (equivalent to around US$38,100 or £28,300 today) on importing ladybirds and releasing them into their orchards, in the hope that an Australian predator might be able to control the Australian scale pest where the local ladybird species had failed.

The farmers' prayers were answered with the Vedalia Beetle. With its ability to disperse and search out prey, combined with the narrow range of suitable prey species and the ladybird's rapid population growth rate (through high fecundity, multiple generations in a year and long-lived adults), the species turned out to be ideal for the pest-control role. It soon began to eat its way through the huge scale insect population, and within two years the citrus trees were bearing harvestable fruit again – the 1890 harvest was three times the size of that in 1889. The ladybird had single-handedly saved an industry, which nowadays is worth US$500 million (£370 million).

The two species coexisted in the California citrus groves for decades afterwards, the ladybird keeping scale numbers low enough such that harvests weren't harmed. However, in the mid-twentieth century great progress was being made with chemical insecticides, and at least two – DDT and malathion – were used in the orchards. The insecticides killed both insect species, but the scales recovered first, and their resurgence caused all the problems last seen in 1889 to resurface. Eventually, common sense dawned, the chemical spraying was reduced, the ladybirds came back and the scales were devoured – again.

The overwhelming success of the Vedalia Beetle in California set off a biocontrol bandwagon that ran roughshod through the late nineteenth and early twentieth centuries, ricocheting from success to failure and back again, until synthetic pesticides largely took over in the middle of the twentieth century. Several of the successes were as comprehensive as the Californian experience. The Vedalia Beetle has since been introduced to more than 40 countries worldwide to control the Cottony Cushion Scale or other scale insects, and in each case substantial or complete control has been achieved.

Several other ladybird species have also been used successfully to control scale insects; members of the genera *Chilocorus*, *Cryptolaemus* and *Rhyzobius* have been

Below: The Cottony Cushion Scale can reach population sizes that suck trees dry unless something eats them first.

Above: The St Helena Gumwood owes its survival to a species of tiny ladybirds – *Hyperaspis pantherina*.

particularly effective. In 1993, the ladybird *Hyperaspis pantherina* was released on St Helena, where the invasive ensign scale insect *Orthezia insignis*, accidentally introduced to the remote island from South America, was steadily wiping out the entire global population of the endemic St Helena Gumwood (*Commidendrum robustum*), the island's national tree. The ladybird successfully controlled the scale insect, and in doing so almost certainly saved the tree from extinction.

Many introduction events did not produce such positive results, however. In 1891–92, following the successful introduction of the Vedalia Beetle, a further 46 ladybird species were sent from Australia to the USA. Unfortunately, very few of these became established and none had the same level of success. In particular, attempts to use ladybirds to control aphids have generally failed; even where the ladybird manages to establish a breeding population, control of the aphid pest has generally remained elusive. This is thought to be for two reasons. First, aphids breed far more quickly than either scale insects or ladybirds, so their populations can increase at such a rate that there are never enough ladybirds to go around. And second, very few aphid-feeding ladybirds are so specialised that they feed only on a narrow range of prey species. This means that most ladybirds, faced with low numbers of the designated pest species, will simply disperse from the targeted area and eat other species instead. This generalisation has led to some catastrophic outcomes. In particular, the Harlequin Ladybird, an Asian species first introduced to the USA in 1916, has had a devastating and far-reaching effect on native species (see pages 86–89).

Other introductions have failed catastrophically but without the same level of impact on other species – aphid or ladybird. As we saw earlier, the Convergent Ladybird assembles in the same sites year after year to spend the winter in massive groups. For more than a century, the species has been collected from these sites, placed in cold storage to prolong overwintering, and sold as aphid biocontrol agents to farmers and gardeners. Even 100 years ago, harvests could exceed 20kg (44lb) – or more than a million ladybirds – per person per day. However, even vast releases of Convergent Ladybirds, numbering millions upon millions of beetles, have failed to control some of their aphid targets – the ladybirds simply flew away almost immediately after they were released. In perhaps the most ridiculous example, in 1990, authorities in the Netherlands, distressed at aphids living in roadside trees and dribbling honeydew onto parked cars, imported around 30 million individual Convergent Ladybirds to combat the problem. The result? The ladybirds promptly flew away and vanished, never to be seen again, and the aphids remained untouched.

Right: Ladybirds are shipped around the world as biological control agents; however, the results can be mixed.

Varied diets

Ladybirds don't just eat aphids and scale insects. Some are real generalists, eating virtually anything smaller than they are, while others have specialised in more unusual prey items. In the UK, the Horseshoe Ladybird (*Clitostethus arcuatus*) feeds almost exclusively on whitefly, while the moorland-dwelling Hieroglyphic Ladybird (*Coccinella hieroglyphica*) mainly eats larvae of the Heather Beetle (*Lochmaea suturalis*). In orchards, the minuscule Dot Ladybird – a hairy full stop with yellow legs – eats mites, particularly spider mites. Further afield, various ladybird species specialise in planthoppers and other true bugs, moth eggs and caterpillars, or even ants.

Over a third of the known ladybird species are not predatory at all, but mainly eat plants or fungi, while most of the primarily predatory species will feed on pollen, nectar or fruit from time to time, especially when aphids or scale insects are scarce. Members of the subfamily Epilachninae feed exclusively on leaves. These species are covered in short, downy hairs, and are generally pale orange with multiple black spots; in the UK, they include the 24-spot and Bryony Ladybirds. The mandibles of both adults and larvae are modified from the single stabbing prong of the typical predatory ladybird into a multi-toothed comb. The ladybirds use this to scrape away the surface of leaves, leaving a characteristic grazed window: the

Left: Most ladybird larvae are smooth and sleek, however, others can be impressively spiny.

Above: As a group, ladybirds have a very mixed diet including aphids and fruit, as well as the plants themselves.

presence of this on False Oat-grass is often the first sign that the 24-spot Ladybird is present at a site.

Most herbivorous ladybirds are harmless grazers of meadow and hedgerow plants, but a handful of species have a taste for crop plants. Normally they consume such a tiny proportion of the crop that there's no effect on the harvest, but in peak years, especially when the parasitic wasps that normally keep their numbers in check are few and far between, the ladybirds can be abundant and problematic. In particular, the Mexican Bean Beetle (*Epilachna varivestis*) and the Squash Beetle (*E. borealis*) are major pests on beans and similar crops in the southern USA in these 'ladybird years'. Native to the Americas between Guatemala and Canada, the 6–7mm (¼in) Mexican Bean Beetle feeds on a wide variety of bean crops as both an adult and larva. Grazing on the lower surface of the leaves, the adult ladybirds and their larvae skeletonise their foodplants, particularly in July and August.

Not all plant-eating ladybirds favour grazing leaves. Several species can survive and reproduce when feeding entirely on pollen, particularly *Bulaea lichatschovi* and several members of the genera *Coleomegilla* and *Micraspis*.

Other species have returned to their ancestral habits of fungus-eating, generally feeding on mildews rather than the more familiar mushrooms and toadstools. In Britain, the Orange Ladybird and tiny (3mm, or 1/8in) lemon-yellow and black 22-spot Ladybird (*Psyllobora vigintiduopunctata*) are fungal specialists, primarily feeding by grazing on the white powdery mildew that coats the leaves of Hogweed (*Heracleum sphondylium*) and other low-growing plants in mid- to late summer. Although the 22-spot is easily overlooked, it can reach very high densities in the right conditions. The 16-spot Ladybird has only partially reverted to life as a fungivore: generally feeding on fungi, it also eats pollen, and gut content analysis has revealed that it is partial to the occasional mite or thrip.

In fact, both pollen and fungi feature far more widely in ladybird diets than was realised until recently. Many species, including the Seven-spot Ladybird, will eat large quantities of pollen and nectar in spring, when aphids are in short supply, and fungal spores, honeydew and nectar are all commonly consumed in autumn, when ladybirds are feeding up before overwintering. These are also the periods when most unusual prey animals are caught and eaten. In summer, when female ladybirds are readying themselves for egg-laying, they tend to hone their diets down to the core set of preferred prey species to maximise nutrition and thus numbers of offspring. But when these primary prey species are scarce, all bets are off and hungry ladybirds will try their luck more widely: they have been recorded scavenging on dead animals, raiding spiderwebs and even drinking human sweat.

Below: Some fungus-eating ladybird species favour mildews, rather than more typical mushrooms and toadstools. Here, this 22-spot Ladybird is grazing on the mildew growing on the surface of a leaf.

Ladybirds in the Landscape

Standing out because of their bright warning colouration, a handful of ladybird species can be found almost anywhere, including parks and gardens. But it requires a bit more searching – in different habitats – if you want to catch up with some of the scarcer species. Grassland, moorland and heath, and deciduous and coniferous trees all have their distinct ladybird faunas. Some species are even more specific and can be found only in ivy clumps, on riverside shingle or on tall vegetation around ponds. Whatever habitat you're in, there are likely to be ladybirds in the vicinity.

Most ladybirds live in rough bands of different heights above the ground. Grassland species such as the 16-spot Ladybird, 24-spot Ladybird and *Rhyzobius litura* are rarely found more than 30cm (1ft) or so off the ground, feeding on and in grasses and low plants.

The herb-layer species extend up to about waist height. This group includes the Adonis and 22-spot Ladybirds, as well as the Seven-spot and 14-spot; nettle and Bramble (*Rubus fruticosus agg.*) patches are classic habitats for this group. Sometimes, these species can be more restricted.

Opposite: The Eyed Ladybird can usually be found on trees which, in addition to grassland, moorland and heath, are home to ladybirds.

Left: The 16-spot Ladybird is a classic rough-grassland species that often eats pollen.

Above: The Hieroglyphic Ladybird is very rarely found away from heather.

Below: Ten-spot Ladybirds are a mostly tree-dwelling species but they sometimes turn up at ground level.

The tiny *Coccidula rufa*, for example, is a species of wet grassland and is rarely, if ever, found away from damp, marshy water meadows and similar habitats. Similarly, the Hieroglyphic and Heather (*Chilocorus bipustulatus*) Ladybirds are found almost exclusively on heathland and moorland, usually on heather.

A third set of ladybirds live in trees. These can be further divided into species of deciduous trees, such as the Cream-spot and Orange Ladybirds, and those of conifers, including the Larch, Eyed and Striped Ladybirds. The massive increase in the number and extent of conifer plantations across the UK in the twentieth century provided lots of new habitat for the latter species, and the Larch Ladybird in particular is now widespread across areas of England where, historically, it would have been absent.

While ladybirds generally have habitat preferences and stick to them, there is some overlap between the groups. Two-spot and 10-spot Ladybirds, for instance, are mainly tree-dwelling species but can also be found on nettles nearby. Within these broad habitat types, an area's suitability for ladybirds is defined by its food sources – it is the presence of their preferred prey types that ladybirds are seeking as they disperse widely across the landscape.

A plague of ladybirds

On occasion, during periods of hot weather, ladybirds become hugely, ridiculously abundant. The last such 'ladybird summer' in the UK was 1976, although there have been smaller and more localised events since, most recently in 2009 and 2010. The summer of 1976 was among the hottest on record at the time, and followed a mild, wet winter. Plants grew well in the warm, moist conditions and supported huge populations of aphids. Ladybird numbers, already high after the long, warm summer of 1975 and suffering little overwintering mortality, exploded. In July and August, huge swarms of newly emerged adult ladybirds appeared all over the country, mostly Seven-spots but also including a range of other species. This was particularly evident in coastal towns, where numbers built up to an estimated 24 billion as the ladybird swarms stopped when they reached the sea. Many towns resorted to sweeping the insects up with brooms. Even aircraft were not immune: one pilot reported flying into a swarm of ladybirds at 450m (1,500ft), where they were so numerous that he was forced to land to clear the canopy and clean out the air intakes.

The ladybird swarms had begun as the beetles' food ran out. The new generation of adults emerged from their pupae to find nothing remaining but drought and withered plants as the crops were harvested around them. Aphid populations crashed, caught between the lack of food and the abundance of predators. Starving, the ladybirds dispersed in all directions and began taste-testing anything they could find – even people. Many starved to death, either during the summer or in the winter that followed as their fat reserves proved insufficient to sustain them. Most of the survivors resorted to cannibalism. Huge numbers drowned in the sea attempting to disperse further: it was estimated that around 640km (400 miles) of the tideline around the coast of southern and eastern Britain was carpeted in dead ladybirds.

Above: Drifts of dead ladybirds can be formed in 'ladybird summers'.

Wetland antics

One of the British ladybirds best adapted to its preferred habitat is the Water Ladybird. This is a species of water meadows, and is often found on Reeds (*Phragmites australis*), reedmace (*Typha latifolia* and *T. angustifolia*) and other vegetation right at the water's edge. It is rather more flattened than the average ladybird so as to be able to squeeze between the flat leaves – this is particularly useful in winter, as it allows the beetle to tuck itself right into the centre of the plant. Water levels tend to rise in winter, so Water Ladybirds generally do not go right down to the base of the plant, instead remaining at least semi-alert to approaching floods and moving higher up the stems if the situation requires it.

Scurrying back and forth on wind-whipped reeds over open water is not the most secure of options, and it is far from unusual to see individual Water Ladybirds fall off a plant. Handily, landing in the water is not the death sentence that it would be for most insects: the Water Ladybird can swim! If an individual falls in, or if water levels rise so much and so rapidly that it is floated off the top of the plant as it submerges, it will splay its legs wide, moving them slightly until it has done approximately two full rotations and spotted the nearest dry land or vegetation. Then it begins to swim actively towards the dry spot. While Water Ladybirds may not be the most efficient of swimmers,

Below: Water Ladybirds begin their adult life a yellowy-buff colour to blend in with the dead reeds.

Left: In spring, the Water Ladybird changes from yellow-buff to a bright orange.

they are plenty good enough to get the job done, and are certainly better at it than any other British species.

In addition to its unusual lifestyle, the Water Ladybird is the only British species that changes colour as an adult. When it emerges from the pupa in late summer, it is a yellow-buff colour with black spots, perfectly camouflaged against the dead reeds in which it will spend the winter, a time of year when potential predators are more desperate and the ladybird has less energy and ability to defend itself. But come spring, the adult ladybird wakes up and begins to lay down a new, bright pinkish-orange pigment in its elytra. As it begins to feed, it becomes better able to defend itself, so it is worthwhile regaining the colours that warn off potential predators. Thereafter, it is brightly coloured and stands out from the background like most other ladybird species.

Changing colour like this is fairly unusual in the animal kingdom and, unlike the Water Ladybird, most species that undergo such a transformation do so deliberately to remain camouflaged by blending in more effectively in a new environment. For example, the Green Shield Bug (*Palomena prasina*) is green in summer, but changes to purple-brown in winter to blend in better with dead leaves. Changing to be brightly coloured, or simply being brightly coloured from the start – as is the case in most other adult ladybirds – tells us that there is a real benefit to standing out from the surroundings.

Below: The Green Shield Bug lives up to its name in summer, before changing to purple-brown over the winter.

Warning signs

Most ladybird species are able to synthesise noxious or toxic organic chemicals called alkaloids for protection against being eaten, but these defences are pointless if the predator has to kill the ladybird to notice them. Therefore, ladybirds – like wasps, bumblebees and strawberries – are brightly coloured to draw attention to themselves and stick in the mind of the animals that eat them. Unlike strawberries, however, which are bright red to attract herbivores to disperse their seeds, the eye-catching colour of ladybirds is a reminder to predators that have tried to eat them once that they taste bad and should be ignored if encountered again. Essentially, the bright colours and patterns of ladybirds are purpose-built to discourage predators from making a meal of them. This phenomenon, known as aposematism, works because predators can learn by experience that certain colours or patterns – like a red ladybird with black spots – are not good to eat and should be avoided. This is also likely to be the reason why so many ladybirds share similar colouration – in the UK, for instance, 15 species are orange or red with black spots. Predators avoiding a species with one pattern end up avoiding a whole range of similar-looking ladybirds, often including those that do not have defences optimised against that particular predator.

Below: Strawberries are bright red to stand out and get eaten; ladybirds are the same colour for the opposite reason.

Left: The yellow fluid produced by ladybirds is reflex blood. It is has a deeply unpleasant taste and therefore works as a defence mechanism.

These warning colours are a first line of defence against attack. Contrary to the common myths that black, or yellow, ladybirds are especially poisonous, toxicity does not relate to the actual colour displayed. However, the concentrations of colour pigments in many species do vary in line with the concentration of the insect's defensive chemicals. This is because both require energy to produce, and so a well-fed ladybird has more resources to make both defensive toxins and pigments to colour the elytra. A bright red ladybird is a bad-tasting ladybird!

Each ladybird is basically a miniature chemical-warfare factory on legs. All the brightly coloured ladybirds that have been tested so far have been able to synthesise an array of toxic or just foul-tasting alkaloids, which circulate in the haemolymph (insect's 'blood'). These alkaloids are bitter-tasting, often pharmacologically active, hydrocarbon chemical compounds, which can be toxic depending on the dose. They are produced as defences by a wide range of other animals, bacteria, plants and fungi; well-known examples include caffeine, nicotine and morphine. The compounds are generally named after the species from which they were first isolated – for example, harmonine is produced by the Harlequin Ladybird (genus *Harmonia*) and coccinelline by the Seven-spot Ladybird (genus *Coccinella*). Around 50 different ladybird alkaloids have been identified, with species producing individual blends of three or so different chemicals.

Below: Bitter tasting alkaloids are widespread and include caffeine, as well as ladybird chemical defences.

These defensive alkaloids are generally made by the ladybird, but a few species accumulate and store chemicals taken from their prey and then use these toxins themselves (often in addition to making some alkaloids of their own). The Seven-spot and 11-spot Ladybirds, for instance, can store alkaloids acquired from their aphid prey. As the aphids themselves acquire the alkaloids from the plants they eat, before storing them and using them defensively, the ladybirds are actually using third-hand chemical weapons alongside their self-made toxins.

The North American *Hyperaspis trifurcata* takes things a step further. This species gains its main defensive weapon, carmic acid, from its preferred prey, cochineal scale insects in the genus *Dactylopius*. The scale insects produce the magenta-coloured acid as an ant deterrent, and the ladybirds use it for the same purpose. Ladybird larvae store the acid for just a short time in their body before use (it breaks down fast and must be constantly replenished), and adults do not store it at all. Humans also harvest the red liquid from the scale insects. It is generally known as cochineal and for centuries it was the only known source of red pigment, with a thriving industry established around farming and harvesting the insects. It's strange to think that the red dye used to colour pictures of the Virgin Mary's cloak – and thus indirectly giving the ladybirds their name (see page 91) – comes from an insect mainly preyed upon by ladybirds, which use that same dye as protection!

Right: Cochineal scale insects produce a red dye (cochineal) that is repurposed by the American ladybird *Hyperaspis trifurcata* for their own protection.

Other defences

If their bright colours don't put off a predator, ladybirds don't simply wait to be chewed apart: all ladybirds can 'leak' haemolymph from joints when disturbed. This 'reflex bleeding' floods the mouth of the predator with unpleasant-tasting chemicals – anyone who has witnessed a dog investigate a ladybird will be able to vouch for its effectiveness. Ladybirds aren't injured and don't die when they reflex-bleed, but it is an extra cost for the individual – at the end of winter, when adults are low on resources, some may lose the ability to reflex-bleed, only to rediscover it when fully fed.

Ladybirds can reflex-bleed as both adults and larvae (and some as pupae), while eggs and pupae both contain a solid dose of alkaloids. The alkaloids are spread throughout the ladybird's body, but reflex blood contains a higher proportion of them, to maximise the chances of a predator spitting the ladybird out after just mouthing it, rather than having to bite down to discover its mistake. Adults produce blobs of reflex blood from their knee joints, while larvae have equivalent glands on their abdomens. Additionally, larvae, pupae and adults all display a range of other anti-predator defences: spikes, spines, and hairs for larvae, scissor-like edges to the abdominal sections for pupae and an ability to tuck

Left: All ladybirds can reflex-bleed as a defence, however, some larvae have extravagant physical protection as well.

in anything vulnerable and clamp down on surfaces for the adult.

As a result, most tend not to be targeted by generalist predators except when there is no alternative. They may occasionally still be eaten – when deep snow renders other food sources inaccessible, for instance. There are reports of Magpies (*Pica pica*) eating ladybirds overwintering on shrubs in this situation, despite clearly being in distress – the birds went as far as drinking pond water in between beetles, as if to dilute the taste. In general, however, a ladybird's main threat is from specialist parasites – or other ladybirds. Despite their well-defended nature, the vast majority of ladybirds will die before breeding. For a stable population, just two offspring need to survive, but some ladybird species will lay more than 2,000 eggs per female.

Not all ladybirds are poisonous, however. For example, the Larch and 24-spot Ladybirds rarely, if ever, seem to contain toxins, and the same is true for many of the so-called inconspicuous ladybirds. These species are generally dull in colour and fairly small, and it is suspected that they have ditched the tactic of spending energy on producing complex defensive chemicals and instead opted to be small and undetectable.

Adult ladybirds also employ behavioural defensive tactics, varying these according to the threat. For example, when small predators such as ants or spiders are nearby, the ladybird will hunker down by tucking its legs and antennae under its body and clamping down on the surface beneath, sometimes deliberately leaning towards the threat to keep that side pressed as tightly as possible to the surface. When larger predators are detected, the ladybird instead sidles to the edge of the leaf or other surface, and either drops off into the undergrowth or flies away. The spots on a ladybird's elytra – particularly the eye-spots (dark spots ringed with pale pigment) present in species such as the Eyed Ladybird – may act as false eyes to draw a predator's attention away from the real head, a ploy used by other insects such as the Peacock Butterfly (*Nymphalis io*).

Below: The Larch Ladybird has opted for camouflage rather than warning colours.

Colour variations

A few ladybird species manage to be both eye-catching and blend in with their surroundings. The Cream-streaked Ladybird, for example, is strikingly bright when seen on the green pine needles of its preferred conifer host trees, being variably orange to salmon pink, with diffuse cream streaks front to back and 4–16 black dots in rows across the elytra. However, when the species is resting it does not sit on the dark green needles; instead, it moves to sit on or around the bright orange-red pine buds, where it blends in perfectly against this lurid background.

Some ladybirds are highly variable, or have many different colour forms within the same species. Both the Harlequin and Two-spot have more than 100 named colour forms, for instance, varying from completely black to completely red. This is thought to be an adaptation to help with one particularly important environmental constraint: temperature. Black areas absorb more energy and thus heat up more quickly than do red or orange areas, and they also heat up more when the light or heat is less strong.

Below: The Cream-streaked Ladybird blends in with shoots and buds, however, it can stand out against the green pine needles. When resting, it will often remain near the buds where it is better camouflaged.

Therefore, darker ladybirds can reach the same body temperature – and thus activity level – when there is less light, or the air temperature is lower.

In cooler areas, and in areas that suffer from smoke pollution, less sunlight energy reaches the ladybirds and so darker colouration is advantageous. This is not for camouflage purposes as with moths, but simply because darker individuals become warmer, and thus more active, in low-sunlight conditions. After the Clean Air Act 1956 came into effect in the UK, large drops were seen in the ratio of darker- to lighter-coloured individuals as pollution decreased and light levels increased.

Right: In Britain, the orange colour form makes up around 80 per cent of the Harlequin Ladybird population.

Ant associations

A few ladybird species live closely with ants, turning predators into unwitting allies. In the UK, two species display this behaviour: *Platynaspis luteorubra* and the Scarce Seven-spot Ladybird. The advantages of living around ants are clear. The ants 'farm' aphids for their honeydew, so there is plenty of food, and there is little competition for this because the ants kill or drive away potential aphid predators. The ants even seem to reduce the level of attacks on ladybirds that live alongside them by preventing their predators and parasites getting too close – for example, in one study, spiders were found to be up to 11 times more abundant 100m (330ft) from Red Wood Ant (*Formica rufa*) nests compared with 10m (33ft) from the nest.

The two British species manage to live with ants in very different ways. *Platynaspis luteorubra* employs chemical camouflage, its larvae producing chemicals that smell like the aphids the ants are tending. Ants encountering a larva will tap it with their antennae, 'read' the chemical signature, and then allow it to proceed without raising the alarm. Unlike the larvae, the ladybird pupae and adults do not produce their own scent. Instead, pupae turn to leftover scent from the shed larval skin, and are covered with long protective hairs. Adults spend much less time in the direct vicinity of ants and either clamp down onto the surface or simply run away if challenged.

The Scarce Seven-spot employs the opposite tactic. It produces ant-repellent chemicals, and simply aims to make sure that ants don't approach it. If they do, the ladybird can use a range of other physical and behavioural tactics, including clamping down so that the ants can't get a grip with their jaws, and simply running or flying away. Apparently, these are rarely used or needed, however. Experiments with ant species not affected by the ladybird's repellent have found that both ladybird adults and larvae completely ignore ant attacks, including bites and acid sprays. It seems likely that the Scarce Seven-spot has almost completely lost its fleeing response because the wood ants attack it so infrequently, and is now almost blasé about the danger.

Above: The Scarce Seven-spot Ladybird is rarely found away from wood ants.

Below: This larval *Platynaspis luteoruba* is in chemical disguise, which enables it to smell like the aphids that the ants encounter.

Natural Ladybird Enemies

Life in the wild is hard, even for animals that are as well defended as ladybirds. Although most opportunistic predators are put off by ladybird warning colours and chemical defences, especially hungry – or naive – individuals will still occasionally try to make a meal out of them. More seriously, there is a suite of parasites and diseases that are specialised for attacking ladybirds – sometimes in the most gruesome fashion – and these can be responsible for sky-high mortality rates in some years. And as if that wasn't enough, ladybirds now face their most dangerous opponent: another ladybird, and one that is entirely at home with cannibalism…

Predators

The formidable suite of chemical, morphological and behavioural defences employed by ladybirds (see previous chapter) do not keep them entirely safe from generalist predators. Although they are not a staple of many diets, the beetles are eaten by a range of birds, mammals, amphibians and spiders, as well as dragonflies, ants and other predatory insects.

Gut analyses of a wide variety of amphibians – particularly frogs and toads – have revealed that ladybirds are a frequent prey item. Many of the predators examined had eaten multiple ladybirds in quick succession, suggesting that the beetles' chemicals were not proving a deterrent against hungry amphibians. Possibly, chemicals that work as a defence against mammals, birds or invertebrates are less effective against other groups, but so far virtually no work has been done on this.

Birds (and other vertebrates) that pick out their food generally reject ladybirds, although this seems to vary

Opposite: Whether or not spiders actually eat ladybirds, they will certainly kill any that blunder into their webs.

Below: Aerial hunters, such as House Martins (*Delichon urbicum*), will catch ladybirds on the wing.

Above: Most birds will reject ladybirds because they are distasteful, however, they may become a suitable option at times of great hunger.

according to need. Studies have found that Great Tits (*Parus major*) in captivity actively avoid having to eat the 10-spot Ladybird, but gut analysis of wild Great Tits has found that the same species made up a large part of the winter diet of the bird. While they may be distasteful, ladybirds presumably become an acceptable option when the alternative is starvation.

Similar reasons are likely to account for perhaps the most intriguing predator of ladybirds – and certainly the largest – the Grizzly Bear (*Ursus arctos horribilis*). This species has been seen feeding on the huge overwintering aggregations of the Convergent Ladybird in North America. The fact that the ladybirds were already at their overwintering sites in large numbers suggests that the bears concerned were close to hibernation themselves, and the lack of other available food is likely to have outweighed the bad taste of the beetles.

Other species seem to eat ladybirds more frequently. Birds that hunt flying insects on the wing and devour their prey whole, such as Swifts (*Apus apus*) and Swallows (*Hirundo rustica*), are unlikely to be able to avoid taking ladybirds and do not show signs of ill effects when they do eat them. Similarly, Larch Ladybirds have been found in the gut contents of Rainbow Trout (*Oncorhynchus mykiss*); presumably they were caught either when flying over or floating on the water's surface. In either case, the predator is unlikely to have had much time to study the ladybird before grabbing it.

Some 'predators', particularly spiders, may kill ladybirds without necessarily eating them. Ladybirds can often be seen strung up in webs spun across insect flight-lines, wrapped up and moved to the edge. Presumably having blundered into the web by accident, the ladybirds were killed almost as a reflex reaction by the spider, but apparently not eaten by any but the hungriest arachnids.

Below: Ladybirds that get near spiders' webs are dicing with death. Although arachnids don't often eat ladybirds, they will kill them instinctively.

Parasitoid wasps and flies

The typical image of a parasite is of a tiny animal hitching a lift – and taking a meal – from a much larger one, without doing the host much harm, like a flea or tick on a pet dog. A large group of parasites, however, are more demanding. Known as parasitoids, these are generally much closer in size to the host and, as a result, the host usually dies through the association.

A wide range of parasitoid wasps and flies attack ladybirds. In the UK, the most common of these, or at least the most frequently seen, is the braconid wasp *Dinocampus coccinellae*. Adults are a couple of millimetres (1/16in) long, and black with striking green eyes. They do not need to mate to reproduce, as all are female and asexual. All they need, in fact, is an adult ladybird to act first as food, and then later as a zombie bodyguard! On average, the wasp kills around 10–30 per cent of each of its favoured host species – the Seven-spot and 11-spot Ladybirds – but in small areas in some years the infestation rate can be over 90 per cent. In addition to the Seven-spot and 11-spot, many other ladybird species are used by the wasps as hosts – the species has been recorded emerging from 22 of the 26 British conspicuous ladybird species, although most are attacked at a far lower rate.

Below: This female *Dinocampus coccinellae* is trying to implant an egg into a Seven-spot Ladybird, using her ovipositor to penetrate the weak spot between abdomen sections.

The female *Dinocampus* wasp is drawn to movement, and will land beside an adult ladybird before inspecting it, then tapping it with her antennae and ovipositor. Once she is happy with her choice of host, she will drive her ovipositor into the ladybird through one of the weak spots between sections of its abdomen and lay a single egg. Each ladybird will provide only enough food for one *Dinocampus* larva, but several wasps may lay eggs in a single ladybird – the record is 47! When the wasp egg hatches, the first-instar larva swims through the body of the ladybird, killing any other *Dinocampus* eggs or larvae it comes across, before settling in the ladybird's fat store and shedding its skin, changing into the sedentary, second-instar larva.

The *Dinocampus* larva, however, is not the only thing to hatch from the egg. While the larva goes on its voyage of destruction, 30–100 spherical cells covered with tiny finger-like protrusions called microvilli are released from the egg, and begin to absorb and store nutrients from the host. This drain of nutrients causes similar changes in the host to the preparation for overwintering, namely the shrinking of the reproductive system and increased use of stored fat. For the rest of its time as a larva, the wasp eats these trophic cells rather than the host's tissues. If attacking a ladybird in late summer, the wasp can overwinter along with the host, before stirring into action in the spring as the ladybird wakes up.

Below: The *Dinocampus coccinellae* larva emerges from the paralysed ladybird (top) before spinning a cocoon between its legs (bottom).

Eventually, the wasp larva is fully grown. At this stage it immobilises its host by attacking the motor neurones controlling the ladybird's legs: paralysed, the ladybird can still jerk, twitch and reflex-bleed, but it can't move from the spot. The wasp larva forces its way out between the ladybird's abdominal plates, before spinning a silken cocoon about the size of a grain of rice between the legs of the host. The ladybird now assumes the role of zombie bodyguard; standing astride the wasp's pupa, its defensive adaptations are used (inadvertently) to protect the parasitoid. After a week or so, the newly adult wasp slices its way out of the pupa and cocoon, leaving the ladybird to succumb to starvation, fungal disease or its injuries.

A variety of other host-specific parasitoids attack ladybirds. The fly *Medina separata* attacks adults in a similar fashion to *Dinocampus coccinellae*, while a variety of other fly and wasp species seek out and parasitise juveniles. One stage that is very rarely attacked is the egg. It seems that the ladybird larva's habit of eating any unhatched eggs makes egg parasitism an untenable strategy, except in a couple of species of wasp that solely parasitise the eggs of herbivorous ladybirds.

In the UK, the pupa is probably the most vulnerable stage during a ladybird's life cycle. Stuck, attached to a leaf out in the open, it is unable to run or fly away; it is just able to flex up and down on the spot. Consequently, a wide range of miniature parasitoid wasps will attack ladybird pupae, including members of the genera *Tetrastichus, Oomyzus, Aprostocetus, Homalotylus, Baryscapus* and *Pediobius*. These are gregarious, unlike the solitary *Dinocampus* and *Medina*, with multiple adults emerging from a single host – as many as 40 or 50 have been recorded emerging from the pupa of a single Seven-spot Ladybird. The most common British parasitoids of ladybird pupae are phorids of the genus *Phalacrotophora*, also known as scuttle-flies. Measuring around 1mm ($^{1}/_{32}$in) in length and pale brown in colour, three species are known here: the rare *Phalacrotophora beuki*, and the common and widespread *P. fasciata* and *P. berolinensis*.

Although ladybird pupae cannot run away, they are not defenceless. The abdominal sections are hard and overlap each other as the pupa flexes up and down, potentially trapping or damaging anything a parasitoid might try to jab between them. This has been found to deter around 25 per cent of attacks; consequently, phorids get sneaky. There's a period of about a day between the time a fully grown ladybird larva attaches itself to a substrate, effectively immobilising itself, and it actually shedding the final larval skin to reveal the pupa, plus another couple of hours before the pupal skin is fully hardened. This is the phorid's window of opportunity. Female phorids seek out immobilised prepupal ladybird larvae and, once they find one, begin emitting pheromones to attract a male fly. Once she has mated, the female phorid waits patiently for the ladybird's moment of pupation. When the new, soft pupa is revealed, the fly immediately lays a batch of eggs on it, usually on the underside of the thorax.

Over the course of a couple of weeks, the resultant phorid larvae eat the ladybird from the inside out, leaving just the skin. Fully grown fly larvae force their way out of the pupal skin through a single hole, usually at the front of the underside of the thorax, just below the head, and drop down to pupate in the soil. The number of flies that can emerge depends on the size of the ladybird, but up to 30 have been recorded from a single Seven-spot pupa.

Below: A phorid fly investigates a prepupal ladybird larva (top). The immobile life stages are when a ladybird is at its most vulnerable (bottom).

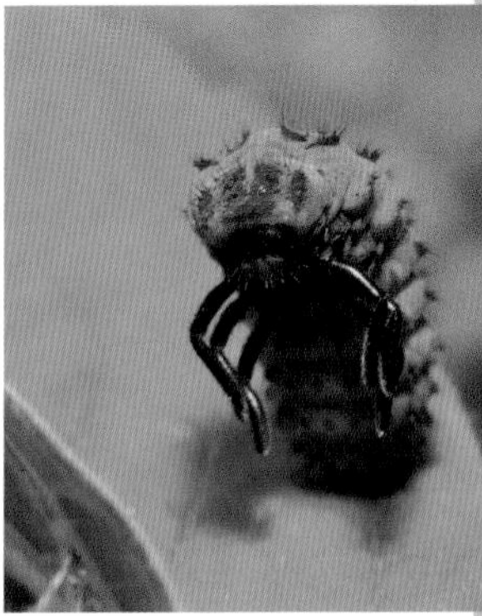

Parasites and diseases

Ladybirds are also affected by parasites and these, along with diseases, can be transmitted during mating. Mites in the genus *Coccipolipus*, found across Europe, generally live under ladybird elytra. They are particularly common on the Two-spot and 10-spot Ladybirds, with up to 90 per cent of some populations of these species infected. The mites live with their mouthparts semi-permanently embedded into the host's elytra or the top of its abdomen, feeding on its haemolymph. Female mites lay eggs that hatch into mobile larvae, and these congregate at the rear end of the ladybird. When the ladybird mates, the juvenile mites swarm across and infest the new ladybird, before shedding their skins to become adults in turn.

The mites' need for overlapping adult generations (they can't survive on ladybird eggs, larvae or pupae, but are reliant on the mating between ladybird adults of consecutive generations) means that they are rare or absent in British ladybird populations, which generally have just one generation per year and virtually no intergenerational matings. Infested female ladybirds rapidly lose the ability to reproduce, laying around 25 per cent fewer eggs. For the eggs that are laid, the hatch rate drops to zero within a matter of weeks.

Another sexually transmitted infection that afflicts ladybirds is caused by a group of fungi known as the Laboulbeniales. These grow as pale yellow finger-like projections on the outer surfaces of the ladybird. They're generally found on the underside and front of males, and the back of the elytra of females – the areas that touch when the two are mating. The fungi don't seem to harm the ladybird noticeably, and are effectively the athlete's foot of the ladybird world!

Below: In addition to predators, ladybirds are also vulnerable to disease. *Hesperomyces virescens* is a common Laboulbeniales fungal disease of ladybirds.

Male-killing bacteria

If ladybirds manage to mate successfully, one further thing can go wrong. Worldwide, at least 14 species of ladybird are known to carry one or more of the five groups of bacteria that kill male offspring as eggs. The bacteria are passed on in the cytoplasm of the egg, but as infected male offspring don't lay eggs and so are incapable of passing the bacteria on, bacteria ending up in a male egg essentially commit suicide. They kill the egg (and thus their own life-support system) to give their near-identical siblings in female eggs a better chance of survival.

This is a successful strategy because hatching ladybird larvae eat any unhatched eggs they can find, including those from the same clutch. If half the eggs – the male half – fail to hatch, each surviving (female) larva has, on average, one egg to eat immediately. This has several advantages. First, it allows the larvae to live longer without finding an aphid to eat. Second, it means that the larvae are bigger when they find an aphid, and are thus better able to attack their prey successfully. More nebulously, the risk of a ladybird larva being eaten by one of its siblings decreases, as there are fewer siblings and more food to go around compared to uninfected egg clutches – again, of benefit to both the female larvae and to their bacterial passengers.

Where bacterial infection rates are high, there can be a marked imbalance in ladybird sex ratios, with males heavily outnumbered by females. Luckily for the survival of the ladybird species, males never seem to disappear completely, because transmission of the bacteria from one generation to the next is never 100 per cent efficient. In at least one species, the tropical Six-spotted Zigzag Ladybird, the transmission can be much less effective. This ladybird has evolved resistance to the male-killing bacteria that infect it in the form of a 'rescue gene'. If either the male or female ladybird carries this gene, any resulting male offspring produced from a mating between the two will be 'rescued', surviving the effects of the bacterial infection.

Below: *Rickettsia* is one of the five male-killing bacteria found in ladybirds (along with *Wolbachia*, *Spiroplasma*, *Flavobacteria* and y-proteobacteria).

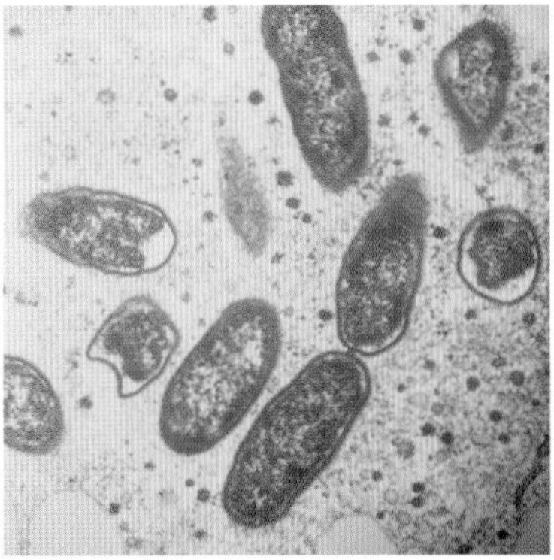

Introduced invaders

Perhaps the single greatest threat to many ladybird populations is the presence of other, more predatory ladybird species. In particular, non-native invasive ladybird species are proving a real problem for native species.

In the USA and Canada, a succession of ladybird species have been deliberately or accidentally introduced since 1880. While most have had no discernible effect, others seemed to be linked to the decline of a range of native species. The Seven-spot Ladybird, native to Europe and much of Asia, was introduced to the USA to control aphids during the 1950s, 1960s and 1970s. None of these deliberate introductions established a population, but a wild population was found for the first time in New Jersey in 1973, apparently from an accidental import. It has since established itself spectacularly well, and competition with the new species is thought to have played a role in the ongoing declines of the Nine-spotted (*Coccinella novemnotata*), 13-spot, Three-banded, Transverse (*C. transversoguttata*), Parenthesis (*Hippodamia parenthesis*), and Convergent Ladybirds across the USA and Canada.

Horrible Harlequins

In the UK, the threat of invasive non-natives is exemplified by the story of the Harlequin Ladybird. Originally found in China, Japan and much of temperate Asia, the species is large (slightly bigger than a Seven-spot) and occurs in a range of colour forms. It was introduced to the USA from 1916 onwards to control aphids, and to France and other European countries – although never the UK – from 1982.

Wild populations of the Harlequin were first found in the US from 1988, and in northern Europe from 2000–01. A handful of individuals were found in Britain in 2003, before the floodgates opened in 2004 and the species rapidly spread across the country. From the genetics of the UK population, the Harlequin seems to have established itself here from at least five separate introductions: individuals from France and Belgium flew across the

Above: The Harlequin Ladybird seems to be doing particularly well in urban areas.

Channel to arrive in southern England, and others were shipped over from North America as accidental passengers on fruit, vegetables and other imports. Between 2004 and 2006, the Harlequin moved 58km (36 miles) northwards and 144km (90 miles) westwards across Britain each year. The spread began to slow after that, but the species is now abundant from Land's End to Edinburgh, with scattered records further north. It reached Ireland in 2007 and,

Left: Although the Seven-spot Ladybird is native to Britain, it can be invasive elsewhere.

Above: Harlequins are highly cannibalistic, and both adults and larvae will eat their kin as well as other ladybird species.

worldwide, the species is now found as a non-native in more than 40 countries.

This is a problem for native species. Each Harlequin eats a lot of aphids, which can reduce the availability of food for native ladybirds – especially larvae, which can't fly to new aphid colonies. Worse, the Harlequin isn't too fussy about what it eats. Even when aphids are around, the species will eat pretty much anything smaller than it is – including the eggs, larvae and pupae of other ladybird species (or other Harlequins). Additionally, Harlequins produce two or even three generations per year in the UK (and more still in warmer countries), whereas virtually all natives only ever produce one. This means that the Harlequin can soon build up huge populations, much faster than native species. Harlequins are also less vulnerable to predators and parasites than natives, and can thrive in virtually any environment. Unfortunately for native species, they are under attack from a near-perfect predator.

Since the Harlequin established itself in the UK in 2004, seven of our eight commonest ladybird species have either begun to decline, or have had an existing decline hastened by the arrival of the new species. Across all British ladybird species, a degree of dietary overlap with the Harlequin and the level of urbanisation within their ranges are the two features that most closely correlate with declines in distribution since 1990.

The adverse effects of the Harlequin can be tracked as the species radiates across the country. The higher the degree of overlap between a native species and the invader in terms of prey types eaten or areas inhabited, the more that species has declined. For example, in the five years following the arrival of the Harlequin in Britain, the Two-spot Ladybird – which eats the same aphids, lives in the same habitats and even overwinters in the same sites as the invader – was lost from 44 per cent of its former range. Even where the Two-spot is still hanging on, it has been driven from being the most abundant species to the threshold of detection. The shipment of the Harlequin Ladybird around the world far outside its native range has unleashed such a voracious predator on native species that they are now struggling to survive.

Below: The native Two-spot Ladybird has suffered huge declines since the Harlequin established itself in Britain.

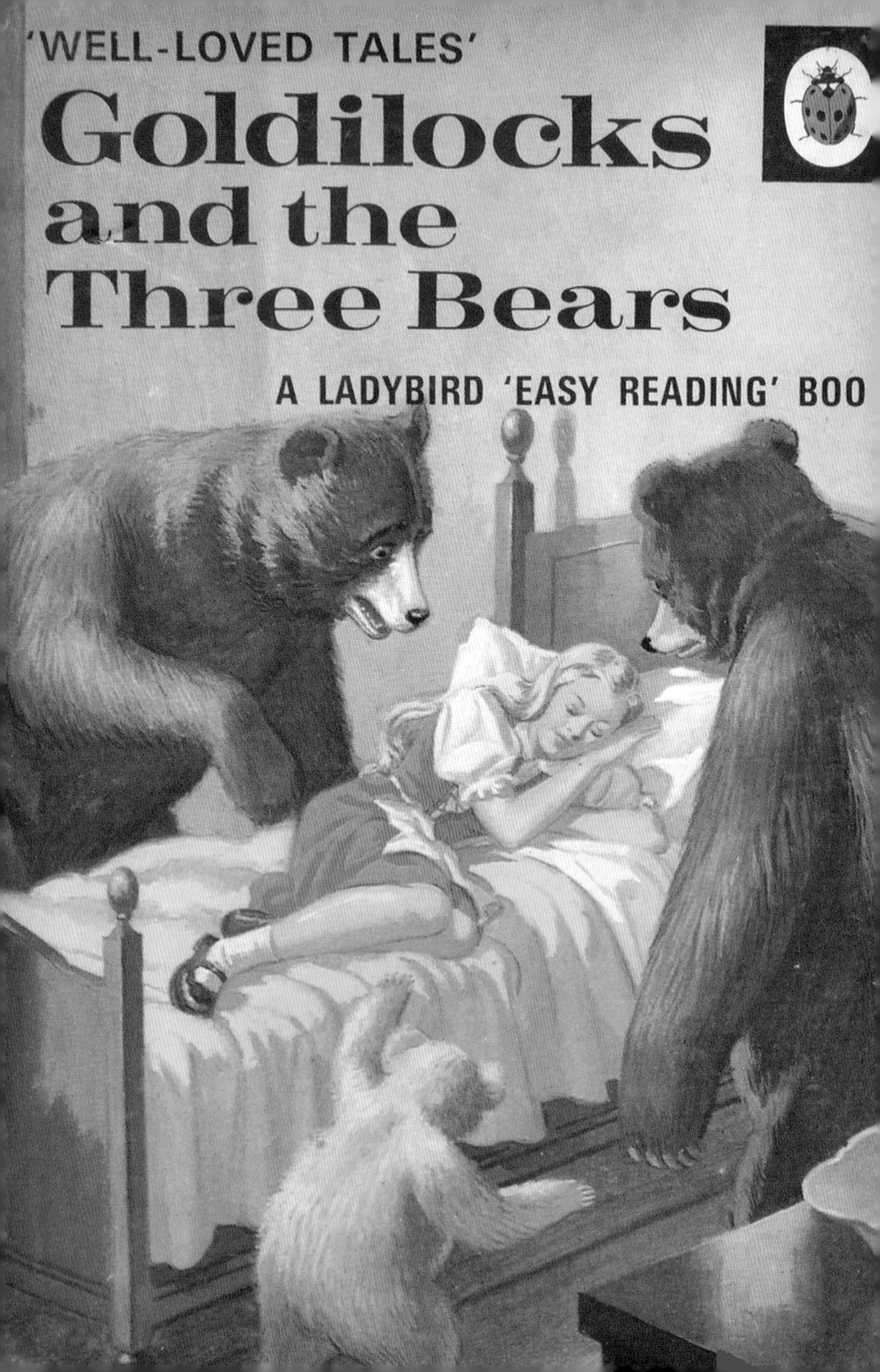
'WELL-LOVED TALES'
Goldilocks and the Three Bears
A LADYBIRD 'EASY READING' BOO

Cultural Connections

As brightly coloured, showy insects that do good in the garden, ladybirds are high-profile beetles. Public interest in them is noted as far back as the fifteenth century, and they feature in a wide range of folk tales and nursery rhymes. More recently, they have lent their image and moniker to causes as diverse as robotics, women-only gyms and a Second World War gunboat. Even the ladybird name itself is linked to myths and legends whose origins date back at least hundreds of years.

What's in a name?

As we have seen, most ladybirds are predators, devouring huge numbers of plant pest insects, and in a roundabout way, this is what led to their ill-fitting name – as ladybirds are neither birds nor are they all female! Since the development of agriculture, farmers have prayed to the gods for deliverance from aphids and other pest insects that devour crops in the field. In the medieval era, they would often have prayed to 'Our Lady', the Virgin Mary, who at the time was usually depicted wearing a red cloak. Shortly after aphids appeared on crops and farmers offered up their prayers, large numbers of small beetles would arrive. Eating only aphids, they were cloaked in red and, as if to confirm matters, they bore seven spots on their backs – a clear allusion to the seven joys and seven sorrows of Mary.

Perceived as a gift from the gods, these tiny predators became known as 'Our Lady's birds'. Over time, this became shortened to 'ladybirds'. In some parts of Scotland ladybirds are still called 'ladyclocks', where 'clock' is a corruption of 'cloak', as worn by Mary. Killing a ladybird was thought to put the culprit in Mary's bad books for nine days, bringing sadness and misfortune, as the insects were said to live under her protection.

Opposite: The 11-spot Ladybird has inspired names and logos across the world, including the beloved Ladybird Books.

Below: *Madonna im Gärtchen*, an early sixteenth-century painting of the Virgin Mary by Matthias Grünewald, depicts her wearing red.

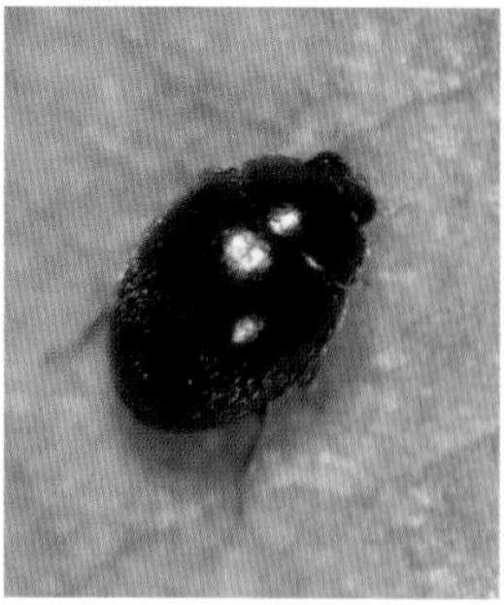

Above: Calling black ladybirds 'the Devil's chickens' contrasts with the folk association between ladybirds and the Divine.

This association was not just made in the English-speaking world. A 1991 study of the folklore associated with ladybirds worldwide found 329 names in 55 different languages. Of these, 25 per cent dedicate the species to the Virgin Mary, including the German *Marienkäfer* (Mary's beetles) and the French *bêtes de la Vierge* (animals of the Virgin), and 15 per cent to other religious figures, including God, Jesus and the Pope. One exception is Italian, where black ladybirds were known as *galineta del Diabolo*, or the Devil's chickens! Nor was the association between ladybirds and the Divine restricted to Christianity. In Norse mythology, Thor sent the ladybird to Earth riding on a bolt of lightning, and some Asian cultures see ladybirds as interpreters for the gods.

Bishop Barnaby

In England, particularly in Norfolk and Suffolk, ladybirds are traditionally known as Bishop Barnaby (or variations on this, including Barnabee, Burnabee, bishy-bishy-Barnabee, and bishop-that-burneth). How this name came about is unclear, but the most likely theories are that it is a contraction of 'bishop-that-burneth', from the fiery red elytra of the common ladybird species, or a perceived connection to St Barnabas, whose feast day falls on 11 June, just as the beetles begin to appear in large numbers. A further possibility is that the name was inspired by an actual bishop, either named Barnaby or of a church of St Barnabas, who wore a red cloak – probably the cappa magna, or great cape, worn by Catholic clergy. However, as there are no records of any Bishop Barnaby in the region, it seems that the true origin of the name will remain lost in the mists of time.

Above: St Barnabas has long been associated with ladybirds, in part because his feast day coincides with the appearance of ladybirds in June.

Legends of the ladybirds

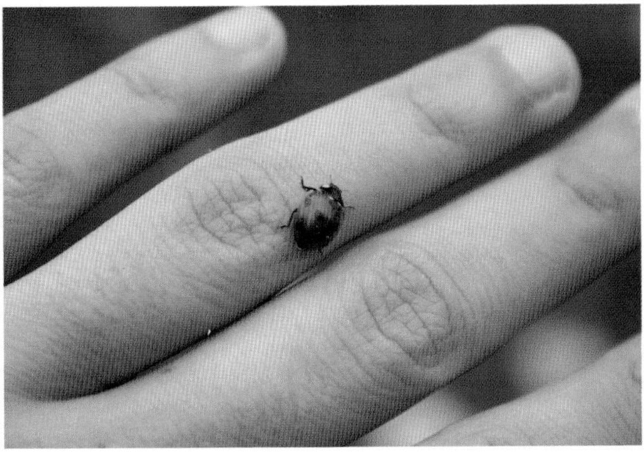

Left: It is often thought that a ladybird landing on someone will supposedly grant them a wish.

Not surprisingly for insects seen as signifiers of the heavens, ladybirds are considered harbingers of a bountiful harvest and good luck in many cultures. In Turkey, the word for ladybird, *uğur böceği*, literally translates as 'good luck bug'. A wide range of traditions and folklore have grown up around this perceived aspect of the insects. For example, when a ladybird lands on someone, their wish will supposedly come true. An alternative of this suggests that whatever the ladybird lands on is soon likely to be upgraded to a new and improved version.

Below: It is believed that the number of spots on a ladybird equals the number of months of good luck that you will have. This Five-spot Ladybird must be one of the more disappointing species to find.

The number of spots on a ladybird is also the subject of folklore. Depending on the particular legend, the number of spots can equal the number of banknotes or coins that will soon come your way, or the months of good luck you will have.

Ladybirds are often considered totems of luck in love. In some Asian cultures, it is believed that if a ladybird is caught and then released, it will fly to your true love and whisper your name in their ear, causing them to come running.

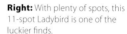

Right: With plenty of spots, this 11-spot Ladybird is one of the luckier finds.

In many European countries, a released ladybird is said to fly off in the direction of your true love – especially if the couplet 'Fly away east or fly away west, And show me where lives the one I like best' is whispered to it before release.

Elsewhere in Europe, the ladybird is considered a harbinger of romance. In Norway, a man and a woman who see a ladybird at the same time are fated to end up together, while across many European countries there is a belief that if a ladybird lands on a young woman's hands, it is measuring her for wedding gloves and she will soon be married. The spots can then be counted and will equal the number of children she will bear in her lifetime. This is reflected in one of the Italian names for ladybirds, *commaruccia*, meaning 'little midwives'. In Switzerland, babies are delivered by ladybirds rather than storks – presumably working together in large groups!

In many northern European cultures, the first ladybird sighting of the year means that spring and summer will not be long in coming. *Frejhöna*, *gullfrigga* and *gullbagge*, Swedish names for ladybirds that are thought to predate Christianity, all mean 'golden beetle', probably due to an association with the sun. Some of the alternative common names for ladybirds reflect this more clearly: to Estonians, ladybirds are *kirilind*, meaning the 'little cow of summer'; and 'sun beetle' is the literal translation of both the German *Sonnekäfer* and Flemish s*onne kever*.

Ladybirds in song

The principal ladybird association in popular culture is with children. For decades, if not centuries, children's toys, books, games and clothes have been adorned with stylised ladybird images, perhaps because the insects' cheerily unsophisticated red and black colouration was bold enough and bright enough to appeal. This relationship is reinforced from the cradle onwards: one of the most famous children's rhymes is all about ladybirds.

Dating from at least 1744, and with a huge number of local variations (a testament to the amount of time it has been passed around by word of mouth, rather than as written text), 'Ladybird, Ladybird' is widely known. One of the most frequently repeated versions goes:

Ladybird, ladybird,
Fly away home,
Your house is on fire
And your children all roam;
All except one,
And her name is Ann,
And she has crept under
The frying pan.

Below: As the poem 'Ladybird, Ladybird' goes, this Seven-spot Ladybird is getting ready to 'fly away home'.

Above: Other versions of the 'Ladybird, Ladybird' rhyme are more sinister. The 'Your house is on fire' line is thought to signify the burning of old Hop vines following the harvest.

Ann (also Nan, Little Anne or other variations) may be sitting in her pan, weaving her laces as fast as she can, or she may be Aileen and hiding under a soup tureen. Whichever version is preferred, they are all thought to originate from one event in the farming year: the burning of old Hop (*Humulus lupulus*) vines after the harvest to kill off pests and diseases, and to clear the land for the next plantings. The rhyme is supposedly meant as a warning to the adult ladybirds that their 'children' are in danger – the mobile larvae can roam, but the immobile 'Ann' is sometimes thought to represent a pupa. A shorter, starker version makes this more explicit:

> Ladybird, ladybird,
> Fly away home,
> Your house is on fire,
> Your children shall burn!

Because of its familiarity, the rhyme (and thus ladybirds) turn up in a range of other contexts, from Mark Twain's *The Adventures of Tom Sawyer* (1876) to the *Hawaii Five-O* TV series, via Beatrix Potter, Ken Loach and a 1963 film about a nuclear attack, usually to evoke childhood or childishness.

Below: Ladybird is often used in the titles of books, TV series and films, including Ken Loach's *Ladybird Ladybird*.

Literary ladybirds

As children learn to read, they may begin with Julia Donaldson and Lydia Monks' *What the Ladybird Heard* picture book series before progressing through the phenomenally successful Ladybird Books, a library of more than 650 titles, the first of which was published in 1914. Intended for children of all ages, each Ladybird book features a prominent logo, a stylised 11-spot Ladybird, which has ensured that this species has been a near-constant presence for generations of children. Nowadays, the beetle even sits atop books written by the Prince of Wales, and Ladybird Book cover artworks can be seen on display in the Museum of English Rural Life at the University of Reading.

Ladybirds aren't solely restricted to children's rhymes and books. The nineteenth-century Northamptonshire poet John Clare, one of the best at celebrating nature, wildlife and the English countryside, wrote at least two poems principally concerned with ladybirds. 'The Lady Flye' chronicles the author's fascination with the insects, while 'Clock-a-clay' was written from the perspective of a ladybird (the 'red, black-spotted clock-a-clay').

Below: The ladybird icon of the popular series of books was ever-present for generations of children.

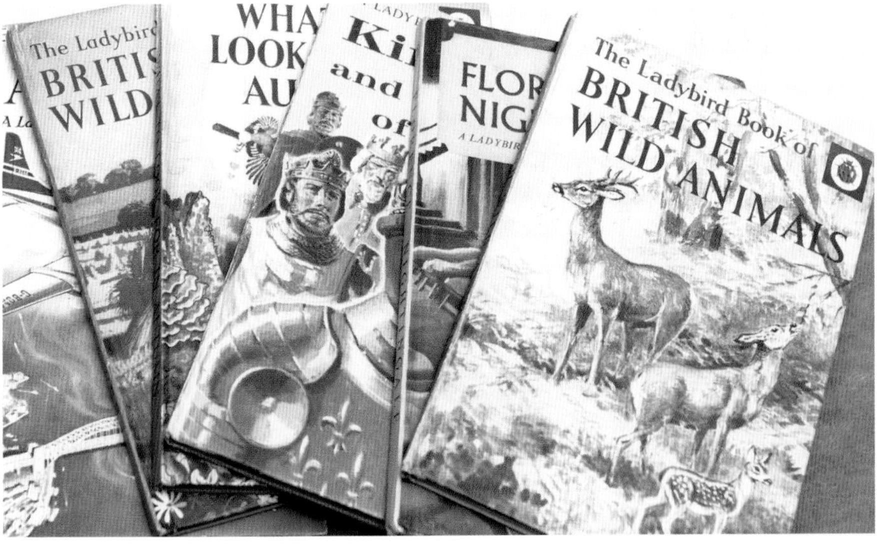

Ladybirds in popular culture

Above: The animated TV series *Miraculous: Tales of Ladybug & Cat Noir* features a character called Marinette who transforms into a ladybird-like superhero by shouting 'Tikki! Spots on!', and fights the supervillain Hawk Moth to prevent him taking over Paris.

Below: Nominated for five Oscars and three BAFTAs, including for Best Picture, Best Director and Best Actress, *Lady Bird* was one of the most acclaimed films of 2017.

Even for those of us who also have non-literary interests, ladybirds are hard to escape. Several amusement parks feature ladybird-related rides, with cars painted up as the red-and-black beetles. In the 1980s, *Pac-Man* was cloned as an arcade game named *Lady Bug*, with similar gameplay but with the *Pac-Man* character replaced by a ladybird. The goal was to eat all the flowers, letters and hearts while avoiding touching the enemy insects that replaced *Pac-Man*'s ghosts.

On TV and at the cinema, ladybirds appear yet again. The 1998 Pixar film *A Bug's Life* features Francis the bad-tempered ladybug, a drag queen who is part of an insect circus troupe. *Lady Bird* (2017) is an American comedy-drama film starring Saoirse Ronan as the eponymous Christine 'Lady Bird' McPherson, a spirited teenager who insists everyone address her by her self-dubbed nickname. In an interview with NPR, writer and director Greta Gerwig explained that after completing her screenplay she recalled the line of the nursery rhyme *Ladybird, Ladybird* 'your house is on fire, and your children are gone'. She believes the rhyme was lodged in her brain as she wrote her coming-of-age story of a high-school senior and her turbulent relationship with her mother. On the small screen, the 1960s TV series *Thunderbirds* occasionally featured a plane called *The Ladybird Jet*. The animated series *Miraculous: Tales of Ladybug & Cat Noir* features the Parisian student and fashion designer Marinette Dupain-Cheng as one of the two main protagonists. With the aid of a pair of earrings infused with the spirit of a magical ladybird-like humanoid, Tikki, Marinette transforms into the superhero Ladybug.

Music gives no respite from the cultural onslaught of ladybird iconography. Between 1962 and 1991, a trio of female singers known as The Ladybirds was a frequent sight and sound on British TV and radio. One of Finland's most influential rock bands is named 22-Pistepirkko ('22-spot Ladybird'). Tadd Dameron's jazz standard 'Lady Bird' is one of the most frequently performed songs in modern jazz, and acts as diverse as Tears for Fears, XTC and Breaking Benjamin have written songs called 'Ladybird' (or 'Ladybug').

Ladybird logos

One of the few brands to survive the 2009 collapse of the Woolworths shopping chain was their children's clothing range: Ladybird clothing. This was originally an independent brand owned by Adolf Pasold & Son that gained its name after company founder Johannes Pasold allegedly saw a ladybird in a dream. At one time the most popular clothing range for the under-fives in Britain, Ladybird garments first appeared in 1938, and for many years adverts for the brand showed ladybirds engaged in human activities. The brand was bought by Shop Direct when Woolworths collapsed, and Ladybird clothing remains a hugely popular line of children's clothing in the UK today.

Above: Adverts for the clothing brand Ladybird generally featured illustrations of the insect as well as children.

Aside from clothing, the classic ladybird image has been used as an icon or logo for a huge range of companies, organisations, projects and products. The Swedish People's Party of Finland has a four-spotted stylised ladybird as its logo, and variations on the theme are used as symbols of, among others, the Croatian National Lottery, a couple of software firms, at least two sororities at American universities, and the ski resort of Candanchú in the Spanish Pyrenees.

In the Netherlands, the Seven-spot Ladybird is used as a symbol of the Dutch Foundation Against Senseless Violence, and tiles featuring a stylised ladybird are set into the street at the sites of deadly crimes. The species has also been adopted as the state insect by five US states (New Hampshire, Tennessee, Ohio, Delaware and Massachusetts). The Nine-spotted Ladybird is the state insect of New York, the only state to pick a native North American species rather than an invasive European import.

Below: Ladybird street tiles in the Netherlands are placed at the site of deadly crimes, as a symbol of the Dutch Foundation Against Senseless Violence.

In Poland, there is an equivalent to Britain's 'Ladybird, Ladybird' rhyme, which (when translated) ends 'Fly to the sky, little ladybird, bring me a piece of bread'. Correspondingly, Poland's largest discount supermarket is named Biedronka ('Ladybird' in Polish) and has a stylised eight-spotted ladybird as its logo.

Ladybirds and technology

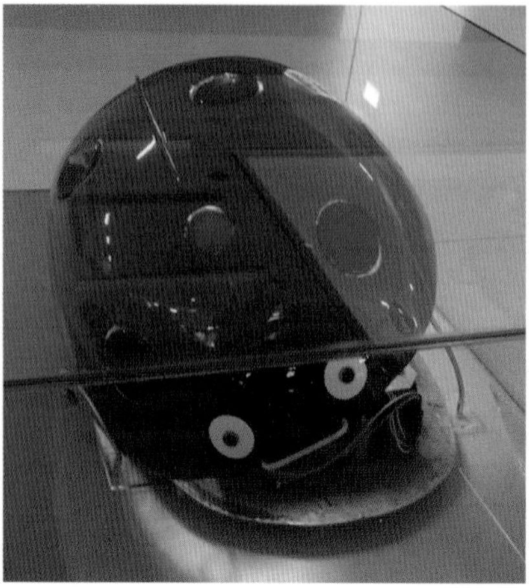

Above: The Ladybird of Szeged, developed in Hungary, was the first robotic animal ever produced.

The first robotic animal ever constructed was a ladybird! The Ladybird of Szeged was developed at the University of Szeged, Hungary, in 1956–57, to mimic Pavlovian conditioning reflexes. Measuring about 60cm (24in) long and 40cm (16in) wide, the robot is a domed ladybird-like shape and has a red shell with black spots, which form part of the control mechanism. The original machine, still in working condition, is now housed in the Informatics History Exhibition in Szeged, where it acts as a mascot.

The robotic ladybird can detect and follow bright lights, and for short periods it can learn to associate the sound of a flute with the presence of bright lights, moving when it hears a flute rather than only when the light is present – at least until its capacitors drain and it 'forgets' the association. When the black spots on its 'elytra' are pressed, the machine enters 'hurt' mode, where it 'forgets' the associations it has learnt and refuses to move.

Ladybirds are also used as the model organisms in biomimetic research, which involves copying and adapting structures perfected by evolution to answer human needs. Ladybirds have complex origami routines to fold away their wings beneath their elytra, and these, along with the physical structure of the wings, have been studied to help build better solar panels for spacecraft.

A handful of ladybirds have even been into space. In 1999, as part of mission STS-93, four ladybirds named John, Ringo, George and Paul were sent into orbit on board the Space Shuttle *Columbia*. Their mission: to boldly explore predator–prey relationships in zero gravity. The ladybirds demonstrated that they were just as able to eat aphids in zero gravity as on Earth, and so moved themselves to the front of the queue to colonise new worlds.

HMS *Ladybird*

Three ships in the Royal Navy have been named HMS *Ladybird*. Two were post-war support ships, but the first was a warship. Launched in 1916, she was an Insect-class gunboat designed for use on the River Danube. The First World War ended before the ship entered service and she first saw action in 1937, when she and her sister ship HMS *Bee* were bombarded by Japanese artillery on the Yangtze River. *Ladybird* was hit six times but not severely damaged, and she was able to steam 30km (20 miles) along the river to rescue survivors from the USS *Panay*, which was sunk by Japanese aircraft.

In 1940, after the outbreak of the Second World War, *Ladybird* was moved to the Mediterranean, where – fitted out with new, larger guns taken from the battleship HMS *Agincourt* – she took part in the North African campaign against the German Afrika Korps. She was used to bombard ground forces and ferry supplies into Tobruk, Libya, but was dive-bombed and sunk on 12 May 1941. The water was shallow enough that the ship did not sink entirely, and her guns were used in an anti-aircraft role for much of the rest of the war.

Above: HMS *Ladybird* at Port Said, Egypt, in 1917, shortly after she was launched.

Left: Four ladybirds – named after members of The Beatles – were sent into space in 1999 on board the Space Shuttle *Columbia*.

Watching Ladybirds

The great beauty of invertebrates is that they can be found almost anywhere, and ladybirds are no exception. With a bit of effort, the amateur entomologist can tick off a decent range of the UK's 47 ladybird species in any garden or park. They can be found at any time of the year, but late spring, summer and early autumn are the best times to search. Another advantage of such showy beetles is that there is no need for special equipment – although some items, such as a hand lens, can improve your experience.

The first warm days in March and April will see Seven-spot Ladybirds basking in sunshine in fresh nettle patches, with more species gradually appearing as the weeks go by. Warm, sunny weather is best; while ladybirds will still be around when it's damp or cold, they won't be as active and are usually tucked down into the foliage, and so are much harder to find.

Location, location, location

The best places to start looking for ladybirds are those where there is plenty of food for them. Lush green plant growth is usually attractive to aphids and thus to ladybirds: nettles, lime trees, Sycamores, birches and various umbellifer species are almost always covered in aphids, and so will also have large resident populations of ladybirds. Scale insects and their predators are often more difficult to find, but Leyland Cypress hedges (*Cupressocyparis leylandii*) – the infamous leylandii – can be surprisingly good for them, as they are often covered in tiny armoured scale insects at the base of the needles. Trees and plants covered in growths of white mildew – particularly Hogweed, Sycamore and oaks – can be irresistible to fungus-feeding ladybirds. Meanwhile, plant-eating species each have their preferred foodplants –

Opposite: Ladybirds can be straightforward to find, however, a magnifier makes them easier to appreciate.

Below: Finding prey, such as aphids, is crucial in being able to detect ladybirds.

Above: When grazing, 24-spot Ladybirds scrape off the top layer of the leaf surface, creating 'windows'.

Above: Ladybird pupae can often be found exposed on leaves.

Above: Ladybird larvae can often be found around colonies of prey insects, such as aphids.

White Bryony hosts the Bryony Ladybird, and Red Clover (*Trifolium pratense*) and False Oat-grass are good places to start your search for the 24-spot Ladybird.

The bright colours of ladybirds make them one of the easier beetle groups to search for by eye, without the need for any complex equipment. Turning over leaves, poking through grass tussocks, and investigating cracks and crevices in bark and buildings, as well as other hidey-holes, can all produce results. One of the easiest species to find by eye is the 24-spot Ladybird. Although this species reaches only about 3mm ($^1/_8$in) long when fully grown, it feeds by scraping off the top layer of the leaf surface, particularly on False Oat-grass. This produces a characteristic pale scar window in the leaf, which becomes very obvious – and common – once you realise what it is. Careful searching in the vicinity of these scars should reveal the dull red adult or the hedgehog-like yellow larva. Their response to disturbance is simply to let go of the plant and drop into the undergrowth, so it may take several attempts before you can study one in detail!

Sometimes, ladybirds will come to you. On a warm day, simply find a sunny spot with some likely vegetation, and sit, watch and wait – the local ladybirds should start to appear. In summer, you are likely to see larvae, and it is fascinating to watch these crocodile-like creatures patrol their chosen patch, remorselessly turning aphids into small, sad heaps of legs and wings. Occasionally, you may be lucky enough to see a batch of eggs hatch, a larva shed its skin to advance an instar, or an adult ladybird emerge from a pupa. These kinds of behaviour, in which ladybirds are interacting with their environment, are best seen by simply watching without disturbing. You don't have to spend long watching a colony to get a real feel for what makes a ladybird tick.

Not all ladybird species are straightforward to find: some species are more inconspicuous thanks to their colouration, habits or simply where they live. To find a wide range of species, it's best to search a variety of habitats, and this means using a couple of extra techniques: sweeping and beating.

Sweeping for ladybirds

For softer, shorter plants such as those found in grassland, meadows and scrub, sweeping gives the best results. This involves sweeping a short-handled net, called a sweep net, back and forth through the vegetation as you walk slowly through a site. Sweep nets can be bought ready-made from entomological suppliers, but you can also make your own fairly cheaply. Bend a thick piece of wire into a 40cm-diameter (16in) circle, then thread it through the hem of a 50–60cm-deep (20–24in) white calico bag. Twist together the ends of the wire, leaving a single twisted 'tail' of 6–10cm (2–4in). Finally, attach the tail end of the wire frame to a 1m-long (3ft) wooden handle with a pair of Jubilee clips or a thick lashing of duct tape.

As you sweep through the vegetation with your net, any ladybirds (and other insects and spiders) in a plant are dislodged when it is whipped out from beneath them, and they drop into the net as it thrashes past. After a couple of sweeps you can examine your catch, either by peering into the bottom of your net or by emptying everything onto a sheet. A cloud of flies and small parasitic wasps will disperse, leaving the bugs and beetles for you to examine –

Below: Sweeping a net through tall grasses will often help locate several ladybird species.

Right: The small species *Coccidula scutellata* is often found at home in wet meadows and damp vegetation.

although these may also fly away if it is a hot day. The key thing is to avoid spiky plants – Brambles and the like will reduce your net to ribbons in short order! Any pots you put your catch into (for no more than about 15 minutes) should be kept out of direct sunlight, and once you've examined your findings, they should be released carefully back where you found them.

Sweeping damp vegetation near water should produce the colour-changing Water Ladybird (yellow-buff in autumn and early spring, then orange-pink in summer), along with the inconspicuous species *Coccidula rufa* and *C. scutellata*. The latter are both a rich, warm red all over, but *C. scutellata* has four black spots on its elytra while *C. rufa* has none. All three species eat aphids from grasses and marginal vegetation around ponds.

Drier grassland will produce the herbivorous 24-spot Ladybird (although to prevent it seeing you coming first and dropping to the ground, you'll need to sweep fast and hold the net as far from your body as possible) and the 16-spot. The tiny, black-spotted beige 16-spot is something of an oddity, as it seems to subsist on a diet of pollen, fungal spores and the occasional aphid, without showing much of a preference for any of them. Also present in grassland in large numbers is the inconspicuous species *Rhyzobius litura*. This is probably the most nondescript of any of the British ladybirds, dull mid-brown all over and around 3mm (1/8in) long. It has a much more typical beetle shape than most other British ladybirds, and is proportionally longer, thinner and less domed than the similarly sized 24-spot.

Sweeping through long grass and mixed vegetation along hedgerows and around recently disturbed areas is a good way to find the fast-moving 11-spot and Adonis Ladybirds. Both prefer warmer sites and reach the northern edge of their ranges in Britain, so sparse vegetation rich in aphids is the best place to come across them – brownfield sites and sand dunes are particularly rich hunting grounds. The Adonis in particular is a striking species, with black spots clustered in the back half of the bright red elytra, as if the ladybird is attempting to outrun them. Where white mildew is present – especially on Hogweed or Meadow Crane's-bill (*Geranium pratense*), the lemon-yellow 22-spot Ladybird is likely to be found, particularly later in the summer or in early autumn.

It requires more strength to sweep through bushy shrubs such as heathers, but these can produce different species – in particular, heathland can harbour the Heather and Hieroglyphic Ladybirds. The Heather Ladybird may be tiny, measuring a mere 2.5–3mm (1/8in) in length, but it is exquisite – a shining, jet-black hemisphere with a row of six tiny, bright red spots on the elytra, which flare out to a lip at the bottom. The Hieroglyphic Ladybird is enormously variable, but the base pattern is a pale orange background with 11 black spots, which vary in size and almost always fuse together into a webbed pattern similar to ancient Egyptian hieroglyphics – hence its name.

Below: Heather and conifers host the Heather Ladybird (left), while the 13-spot Ladybird is a rare species found on wetlands (right).

Beating for beetles

Above: The Seven-spot Ladybird is most commonly found exposed on leaves.

Below: The Cream-spot Ladybird is usually restricted to deciduous trees.

For taller, stronger vegetation such as trees, shrubs and reeds, as well as spiky plants such as gorse or Brambles, beating is the go-to technique. This involves hitting a branch with a stick to dislodge any ladybirds (and other invertebrates) onto a sheet placed below. The sheet is sometimes replaced by a beating tray (a white piece of fabric stretched between a rigid frame), or an upturned white umbrella as a low-cost option.

Beating the low branches of deciduous trees and shrubs will produce the Two-spot and 10-spot Ladybirds, probably our most diversely patterned native species. The Two-spot is usually orange-red with two round black spots, but these can change size and shape, and there are also dark colour forms with a black background and two, four or six red spots across the elytra. The 10-spot is usually pale orange with 10 black spots, but some of these can be very weakly marked or absent altogether – it's reasonably common to find a completely unspotted individual! This species also has a colour form with chequered black and orange markings, and a third form has a purple-black background and a triangular yellow-orange mark on the shoulder of each elytron.

Above: A beating tray is the best way to find ladybirds on trees and tall shrubs.

Present in smaller numbers is the Cream-spot Ladybird, warm maroon in colour and with neatly ordered rows of 14 white spots. The similar Orange Ladybird is found on Sycamores and oaks; it is brighter in colour than the Cream-spot and has 16 white spots arranged in triangles across the elytra. This is a relatively late-season species; as a fungus-eater, it is happier in the warm, wet days of autumn than in the hot, dry summer. Aphid-infested lime and Sycamore trees are beacons for the local Harlequin population, and you may find hundreds at a time!

Coniferous trees are home to a huge diversity of ladybirds. If you can find any, trees with puffs of white waxy filaments on their branches are ideal – the filaments are produced by species of woolly aphid, which ladybirds find very tasty. The Eyed and Striped Ladybirds, Britain's largest species, can be found on needled conifers, generally in small numbers. The larvae of the Eyed Ladybird in particular are spectacular, measuring more than a centimetre (3/8in) in length and very spiny, with bright flashes of orange. By contrast, one of Britain's smallest ladybird species can also be found on conifers, often the same branches. This is *Scymnus suturalis*, 2mm ($^1/_{16}$in) long, hairy all over, and black in colour with large brown patches on the elytra.

Larch Ladybirds (plain brown with occasional cream stripes or a pinkish suffusion) can be found in a wide range of conifers, while the omnipresent Harlequins are likely to be supplemented by the similar, closely related Cream-streaked Ladybird. This is slightly smaller and flatter than the Harlequin, with either four or 16 black spots on an orange background that is streaked with cream. Like the Harlequin, it is a coloniser, with the first British record in 1934, but it is far less of a generalist than its sister species and seems to present no threat to native species.

Even conifer hedges such as leylandii are worth beating for ladybirds, especially trees that look slightly sickly. These often have infestations of scale insects on the twigs, hidden by the foliage, and are good places to find Kidney-spot Ladybirds. These are the bigger brothers of the Heather Ladybird, and are similarly jet black all over but with two large kidney-shaped red marks on the elytra, as well as a red underside to the abdomen. You may also be lucky enough to find a colony of the newly arrived *Rhyzobius lophanthae*, which is very fond of leylandii. This species, native to Australia but widely used as a biocontrol agent

Below: Although *Scymnus suturalis* (left) is tiny, this species is abundant on pine trees. The pupae and adults of Kidney-spot Ladybirds (right) are often found on conifers.

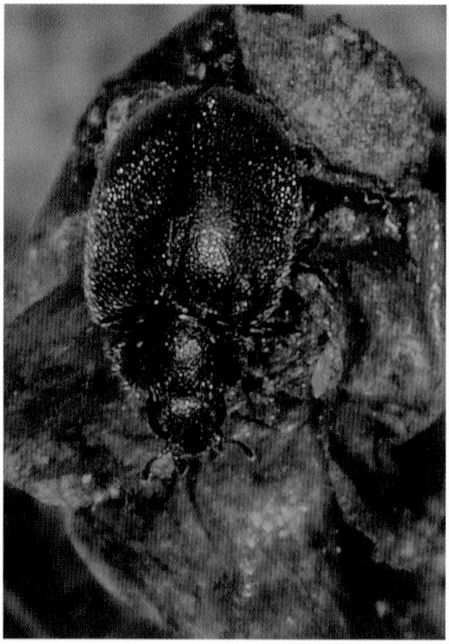

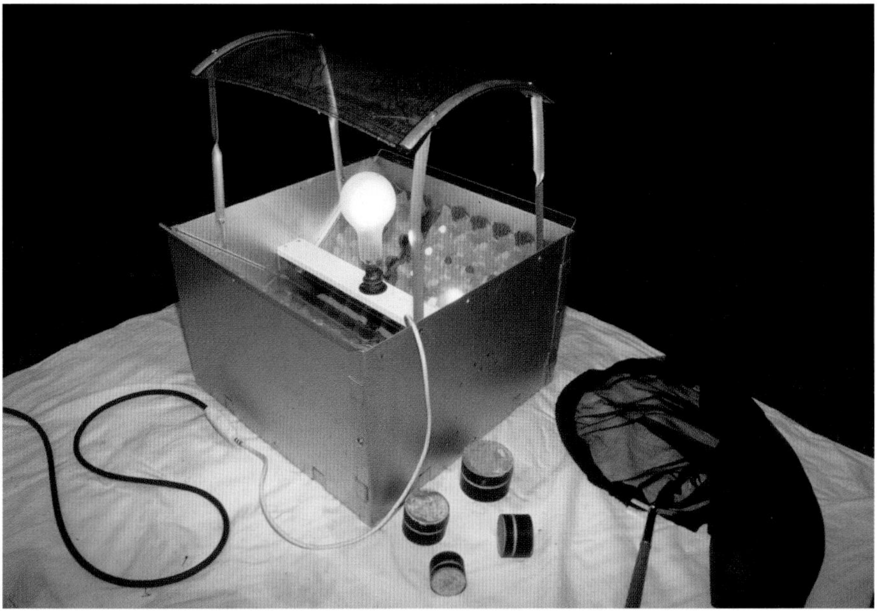

Above: Most ladybirds are attracted to light to some extent, and Orange Ladybirds are frequent visitors.

elsewhere, was first found in Britain in 1999 and has gradually spread across southern England. It is another small species, around 2.5mm (1/8in) long and densely hairy, but with the classic domed ladybird shape. It has slightly reflective gunmetal-grey elytra and a rust-red head and thorax. The larvae are pale grey, with a characteristic rectangular cream marking on the back.

Some ladybirds – in particular, the Orange Ladybird – are attracted to ultraviolet light, and a range of species can be found around outdoor lights or in moth traps. However, these seem to lure only occasional visitors and are not reliable for attracting ladybirds in any great numbers.

When you find a ladybird, it's important to record the sighting with the national recording scheme, the UK Ladybird Survey (see page 125). This way, scientists can use your record to check how populations are changing and monitor things like the spread and effect of the Harlequin Ladybird.

Looking After Ladybirds

Despite their undisputed status as the most iconic of insects, ladybirds have been paid little heed by the conservation movement in general. This is common across most invertebrates, of course, but ladybirds have some crucial advantages when it comes to capturing the public's attention: people generally like them, they are useful to have around and, thanks to the citizen science of the UK Ladybird Survey, we know enough about their former and current distributions to be able to detect population changes.

It is clear that many British ladybirds are in need of conservation, and that some species are already rare and restricted to just a few locations. The Five-spot Ladybird, for example, is found in the gravel and shingle beaches bordering only a handful of rivers in and near Wales (principally the Wye and Ystwyth), and the River Spey in Scotland. The Scarce Seven-spot Ladybird has a scattered distribution, restricted to aphid colonies guarded by Red Wood Ants on heathland, mostly in the south of England.

Opposite: Some ladybirds are familiar sights in towns, gardens and parks, whereas others are much rarer.

Left: Shingle beaches that border rivers are home to the rare Five-spot Ladybird.

The 13-spot Ladybird was one of the first species recorded in Britain, with sightings from the early nineteenth century, but it wasn't confirmed as breeding here until 2011, when the first larva was found at Axminster in Devon. That colony was on a nature reserve, but despite its protected status, the breeding site was submerged beneath floodwaters for most of the monsoon-like summer of 2012 and the ladybird has not been found there again since.

Other species are more widespread but have declined severely. As mentioned on page 89, the Harlequin Ladybird triggered reductions (or more severe declines) in seven of the eight commonest ladybird species across Britain in the first five years following its arrival on our shores in 2004. The 2011 ladybird atlas (*Ladybirds (Coccinellidae) of Britain and Ireland*, by Helen Roy, Peter Brown, Robert Frost and Remy Poland) looked back further and found that 11 species had declined significantly between 1990 and 2010, with just five species increasing significantly (two of which were new colonists during that period).

More recent research found that the two factors that best correlated with declines in distribution across the 25 larger ladybird species were the degree of dietary overlap with the Harlequin and the level of urbanisation. Although ladybirds are often thought of as quintessential urban garden wildlife, only a handful of species can survive in the typical garden of lawn and bedding plants. More specialised ladybirds – species of trees or pond-edge vegetation, for example – can survive in gardens, but only in the right kinds, which are comparatively rare.

Below: This is the first 13-spot Ladybird larva ever found in Britain (left) and the adult that it turned into (right).

Gardening for ladybirds

Above: Aphids specialised on one plant species can be confined, providing plenty of food for ladybirds.

Fortunately, it doesn't take too much hard work to include ladybirds in your gardening plans. In fact, having a ladybird-friendly garden can mean less work – both in the preparation (less tidying) and afterwards (fewer problems with greenfly, blackfly and other pest insects). This allows you to minimise the use of pesticides in your garden, which in turn will allow ladybirds to survive there longer (pesticides kill pests and beneficial insects alike). As pest populations tend to recover more quickly than do ladybirds, lacewings or other beneficial insects, the use of pesticides produces a boom-and-bust pest population, which can peak at high numbers. By contrast, if there is a ready supply of pest-munching ladybirds waiting in the garden, the pests never get away from their predators to the same extent, and outbreaks aren't nearly as bad.

There are two main focus areas to attracting ladybirds to the garden: overwintering sites and food. For most ladybirds, providing overwintering sites simply means leaving the garden in a slightly untidy state during the winter. Let dead leaves gather around the base of perennial plants until the spring growth begins to show, and wait until spring before tidying up herbaceous borders.

Above: Many ladybird species like to tuck themselves away during the winter months.

Generally, leave ground-cover plants alone as much as possible during the winter – those dead or dormant leaves provide excellent shelter for a range of ladybird species.

Evergreen plant species are also particularly useful as ladybird overwintering sites. Pampas (*Cortaderia selloana*) and other grasses form tight tussocks that are great for ladybirds and a wide range of other insects to take shelter in. Evergreen trees and shrubs – including Holly (*Ilex aquifolium*), Holm Oak (*Quercus ilex*) and conifers – provide sheltered sites for species that prefer to overwinter higher up. Orange Ladybirds seem especially keen on Holly and will form small groups all over some specimens, while Box (*Buxus sempervirens*) hedges will shelter large numbers of Seven-spot and 14-spot Ladybirds. Ivy (*Hedera helix*) and, to a lesser extent, other climbers are ideal when they grow tightly around a trunk, branch or wall.

It is easy to be tempted by the wide variety of 'insect hotels' that are sold at garden centres and online, usually with ladybirds listed as a target group. These can work, but they never seem to be a raging success and the same,

or better, results can usually be had by tying together a bunch of twigs and leaving them at the back of the flowerbed.

Once spring arrives and ladybirds begin to emerge from their winter dormancy, their priority is to find food, and they will leave your garden if there is none to be had. The best way to ensure that there is sufficient food for ladybirds is to grow a wide variety of plants – ideally native species as these tend to support more native ladybirds than do non-native plants. It seems counter-intuitive that you can decrease the number of aphids on your prize plants by increasing the total number of aphids in your garden, but, crucially, many aphids are very particular about the plant species they eat. Plenty of plants – including the Common Nettle (*Urtica dioica*), Sycamore and other trees in the genus *Acer*, birches, Red Valerian (*Centranthus ruber*), Yarrow (*Achillea millefolium*) and conifers – support aphids that will feed on them and virtually nothing else (and tend not to suffer too much from the infestation). These aphids will then feed ladybirds without spreading beyond the semi-sacrificial plantings.

Above: In early spring, flowers provide a great place for ladybirds to rest and bask in the sunshine. They can also provide pollen to eat.

Below: Exposed leaves provide basking sites for ladybirds in spring.

Above: Fourteen-spot Ladybirds are voracious hunters of aphids.

In late summer, as aphid colonies wane, many ladybirds top up their diets with nectar and pollen. Making sure your garden plants flower in August and into September will therefore help to bridge the gap between aphid-eating and overwintering. The Orange Ball Tree (*Buddleja globosa*) is fantastic for this – in addition to butterflies and bees, several species of ladybird visit the flowers. If you cut off the dead flower heads every two or three days, the bush should flower for longer. Ivy is excellent as a follow-up: it produces copious quantities of nectar from August right through to November, feeding bees, butterflies, ladybirds and a wide range of other pollinator species. Later on, birds will eat the Ivy berries as ladybirds slumber alongside its stems.

Having a selection of these plants in your garden will ensure that ladybirds are present more or less year-round. And as ladybirds are largely generalist predators, they will keep a beady eye out for other food items in the general area, especially when they're looking to lay eggs and are seeking out smaller, less ladybird-trampled aphid colonies. The aphids trying to sneak over the fence to colonise your prize roses won't stand a chance against the hungry hordes.

Below: Pollen and nectar from flowers can provide a boost for ladybirds at the end of summer.

Special garden features

For those with more space in their garden, or who want to provide habitats for as many ladybird species as possible, larger-scale features can be useful. A pond is one of the best ways to encourage wildlife of all types to the garden, and ladybirds are no exception. Marginal vegetation – especially including rushes, sedges and reeds – is ideal for the Water Ladybird and two of the more attractive inconspicuous species, *Coccidula rufa* and *C. scutellata*.

The 16-spot Ladybird is particularly keen on damp grassland, and both it and the 24-spot Ladybird (along with inconspicuous species such as *Rhyzobius litura* and *Hyperaspis pseudopustulata*) will turn up in patches of lawn left to grow long, especially where these are in sunny sites and adjacent to a pond. Leave the grass to grow, except for one mid-season cut, and remove the clippings. Ideally, mow only half the lawn at a time, cutting the other half a couple of weeks later to give the first half time to recover.

Bare earth and early-succession areas with sparse plants are good for warmth-loving species such as the 11-spot

Above: Garden ponds can provide a suitable habitat for a range of ladybird species.

Below: Bare patches of grass are great for warmth-loving ladybird species.

and Adonis Ladybirds. They will turn up in your borders if these have bare patches and aphids, or around the edges of ponds as the water recedes in midsummer. The ladybirds can be encouraged by digging over a patch of ground and leaving it to regenerate naturally, especially if this is an infertile patch and the more vigorous plant species such as docks are kept in check.

A variety of ladybird-attracting tricks of dubious efficacy can be found with some quick internet research, including luring adults into your garden by providing a mixture of honey, water and brewer's yeast. However, the key thing (which actually works!) is to provide a diversity of plants and habitats throughout the year. Keep things slightly untidy and, crucially, watch and learn. Note which ladybird species are present, and what they're feeding on, breeding on and using in general – and provide more of these. Learn what the eggs, larvae and pupae look like, and teach your family so that they don't destroy them accidentally. Every garden is different, and it will take time to make yours the best ladybird garden it can be.

Ladybird bites and stings?

Ladybirds don't have stings, and are adapted to feeding on small, soft insects. A ladybird walking on sensitive skin can feel 'prickly', like touching a stubbly chin, but this is because each leg ends in a tiny pair of claws.

Ladybirds generally have no need or desire to bite humans, but when desperately hungry or thirsty they will nibble anything and everything, just in case they find something that is palatable. They have small, relatively weak jaws and so find it very difficult to break through human skin, and they will stop when they discover that we're not tasty. However, bites do leave a minute quantity of residue, and in exceptionally rare cases this can lead to an allergic reaction. If this happens, medical attention should be sought immediately.

Tabloid newspapers have reported a few extremely nasty cases of blood poisoning and other serious illnesses supposedly sparked by ladybird bites (almost always blamed on the Harlequin). Rather than any venomous property of ladybird bites, these are testament to the fact that any break in the skin has the potential to become infected.

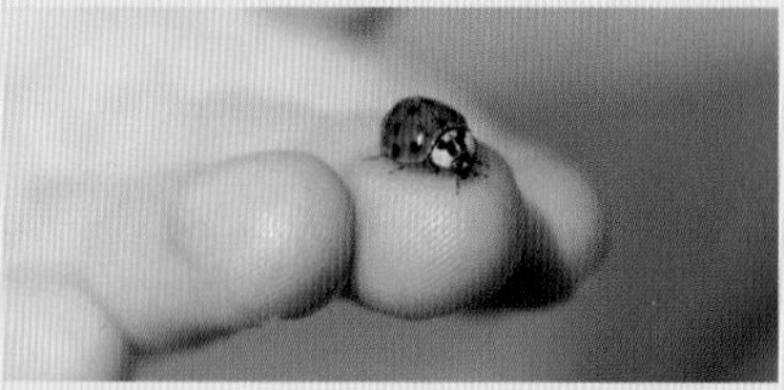

Above: This ladybird's feet may tickle, but it's not dangerous to humans.

Beyond the garden

Along with bumblebees, ladybirds are key to growing food in the countryside. It's estimated that ladybirds are worth more than £65 million per year to British agriculture, simply by eating aphids and other crop pests. A long-term experiment in Switzerland found that, contrary to expectations, crop fields with good conditions for early-spring aphid population growth and lots of nearby aphid overwintering habitat actually had low aphid numbers during the summer. It turned out that aphid numbers ramped up in the early spring in the hedgerows, ditches, wide field margins and other non-crop habitats, which in turn attracted (and fed) large numbers of parasites and predators such as ladybirds. These predators followed the aphids into the crop fields in sufficient numbers to stop the aphid colonies becoming huge, so the overall damage was low.

By contrast, fewer aphids were recorded in spring in fields without non-crop areas, because they had to travel further to get to them. But the few that did make it were able to establish massive colonies, because very few predators travelled that far from the non-crop habitats. So, the aphids ate and ate and reproduced and reproduced, ramping up their population to such a degree that the crops were damaged far more than in fields with large numbers of ladybirds.

Below: A mosaic of fields can provide a habitat for ladybirds, as well as ensuring pest-control for farmers.

Opposite: It is hoped that encouraging the provision of hay meadows will help boost ladybird populations.

One of the key habitat types that prevented crop damage in the study was hay meadows, as they provide a wide range of plant species and thus harbour a wide range of insects. It will be interesting to see how the ongoing practice of encouraging the provision of hay meadows for pollinating insects will affect ladybird populations.

Although many British species are declining, ladybird conservation is not all doom and gloom. The past two decades have been an unprecedented time for ladybird colonisations. In addition to the unwelcome 2003 arrival and establishment of the Harlequin Ladybird, several more benign species have turned up. The herbivorous Bryony Ladybird has colonised the south-east of England, and the 13-spot Ladybird has arrived and bred in both Devon and East Sussex in recent years; time will tell if the latter has established itself here for the long term. Among the inconspicuous species, the list of recent arrivals is even more impressive. *Rhyzobius chrysomeloides* was first found in the south-east in 1997; it is now widespread across England. *Rhyzobius lophanthae* was not a British species 20 years ago; it can now be found in conifers across the south and in the Midlands. *Scymnus interruptus* and *Rhyzobius forestieri* have both colonised our shores, *Cryptolaemus montrouzieri* has been found well away from any deliberate introductions, and new sites (in new counties) have been populated by *Nephus quadrimaculatus*, the Horseshoe Ladybird and *Platynaspis luteorubra*. There is more to do – but all is not yet lost, not by a long way.

Right: The Bryony Ladybird is a welcome new addition to the UK fauna.

Below: *Rhyzobius lophanthae* is originally from Australia, however, it has now made a home in Britain.

Glossary

Abdomen The third section of a ladybird's body after the head and thorax, containing the digestive and reproductive organs.

Aestivation A period of dormancy during dry or hot conditions.

Antenna (pl. antennae) Long, tubular sensory organ on the head.

Aphid Sap-sucking insect, including greenfly and blackfly, forming a large part of the diet of most UK ladybirds.

Aposematism The use of bright colouration to warn potential predators that an organism is toxic, distasteful or otherwise unpalatable.

Biological control (biocontrol) The use of a pest organism's natural enemies (predators, parasites and diseases, collectively known as biocontrol agents) to reduce populations of the pest.

Chitin Carbohydrate polymer made of modified glucose that forms the bulk of an insect's exoskeleton.

Compound eye An eye made up of many individual light-receptor units, as in insects and crustaceans.

Dormancy A period of minimal metabolic activity, entered into to conserve energy.

Elytron (pl. elytra) The hardened wing case of a beetle, protecting the flight wings.

Exoskeleton The rigid external skeleton typical of insects and several other invertebrate groups.

Foodplant A particular plant that is eaten by a particular species, e.g. the foodplant of the Bryony Ladybird is White Bryony.

Generalist A species that eats a variety of different types of food, e.g. the Harlequin Ladybird.

Habitat A particular type of natural environment, characterised by its geography, climate and vegetation.

Haemolymph The insect equivalent of blood.

Hibernation A period of dormancy during cold weather.

Instar The larval stages between moults. The larva is in the first instar after hatching from the egg, but before shedding its skin for the first time (when it enters the second instar).

Larva (pl. larvae) The juvenile stage of an insect, between the egg and the pupa. Generally the main feeding stage.

Mandibles Paired appendages forming part of an insect's mouthparts, used for cutting, biting and gripping food.

Niche The role and placement of a species in its environment. It can be defined by role (e.g. a nocturnal predator), or by preferred or required habitat and environmental conditions (dry heathland etc.).

Overwinter To live through the winter.

Palp Paired elongated, segmented appendages near an insect or crustacean's mouth. Used for feeding, especially tasting and touching food.

Parasite An organism that lives on or in another species (called the host), and that obtains nutrition from the host while providing no benefits to it.

Parasitoid A type of parasite that lives in or on a host, and that kills the host.

Pheromone A chemical released into the environment by an organism to trigger a particular behavioural response in others of the same species.

Predator An animal that preys on others for food, as in ladybirds preying on aphids.

Pupa (pl. pupae) The juvenile stage of an insect between larva and adult, during which the insect becomes mature.

Pupation The act of moulting from a larva to a pupa.

Reflex bleeding Defensive behaviour whereby an insect deliberately exudes a toxic or distasteful fluid – usually haemolymph – when attacked.

Scale insect A small, flattened insect in the order Hemiptera that feeds on plant sap.

Speciation The formation of a new species.

Tarsus (pl. tarsi) The 'foot' of an insect, made up of two to five tarsal segments and ending in one or two tarsal claws.

Thorax The second section of a ladybird's body after the head, mainly full of muscles and concerned with movement.

Further Reading and Resources

The UK Ladybird Survey is the national recording scheme for both conspicuous and inconspicuous ladybird species in the UK. Its websites (coleoptera.org.uk/coccinellidae/home and ladybird-survey.org) include a range of information on ladybirds generally, their identification, and any targeted surveys you might be able to get involved with. Ladybird sightings should be sent through iRecord (brc.ac.uk/irecord).

Books

Recent years have been good for ladybird book publishing.

Roy, H., Brown, P., Frost, R. and Poland, R., *Ladybirds (Coccinellidae) of Britain and Ireland* (Centre for Ecology & Hydrology, 2011).
The results of four decades of biological recording of ladybirds are summarised in this book. It is a distribution atlas for all the UK's ladybird species.

Roy, H., Brown, P., Comont, R., Poland, R. and Sloggett, J., *Ladybirds* (Second Edition) Naturalist's Handbooks series (Pelagic Publishing, 2013).
Covers field- and microscope-based keys to adults of all the UK species and larvae, as well as providing a range of other useful information.

Roy, H., Brown. P (Authors) and Lewington, R. (Illustrator), *Field Guide to the Ladybirds of Great Britain and Ireland* (Bloomsbury, 2018).
This is a useful addition to any ladybird-seeking excursion!

If portability is a concern, the two Field Studies Council fold-out identification charts to ladybird adults (2006) and larvae (2012) are very good, although they cover only the conspicuous species.

For more information away from identification, Michael Majerus's classic *Ladybirds* in the New Naturalist series (HarperCollins, 1994) is very readable and still contains a lot of interesting information, although inevitably there have been considerable advances in many areas in the years since its publication. Majerus's *A Natural History of Ladybird Beetles*, edited by Helen Roy and Peter Brown (Cambridge University Press, 2016), is a more up-to-date look at these fascinating insects.

Ladybird-watching equipment

Watkins & Doncaster has an online catalogue including more or less everything you need to catch and examine ladybirds, from collection pots, sweep nets and beating trays, right up to microscopes (watdon.co.uk).

Anglian Lepidopterist Supplies carries a wide range of ladybird-catching gear, including nets, pots, hand lenses and much more (angleps.com).

Plants

British Wild Flower Plants stocks a very wide range of plants suitable for sowing insect-attracting mini meadows and the like (wildflowers.co.uk).

Emorsgate Seeds sells British-sourced seeds of native wild flowers (wildseed.co.uk).

Flora Locale has a list of native seed and wild plant suppliers (floralocale.org).

Really Wild Flowers stocks plants, bulbs and seeds of native wild flowers, including shrubs and hedge trees (reallywildflowers.co.uk).

Ladybird Species in Britain

Conspicuous or true ladybirds

Two-spot Ladybird *Adalia bipunctata*
10-spot Ladybird *Adalia decempunctata*
Eyed Ladybird *Anatis ocellata*
Water Ladybird *Anisosticta novemdecimpunctata*
Larch Ladybird *Aphidecta obliterata*
Cream-spot Ladybird *Calvia quattuordecimguttata*
Heather Ladybird *Chilocorus bipustulatus*
Kidney-spot Ladybird *Chilocorus renipustulatus*
Hieroglyphic Ladybird *Coccinella hieroglyphica*
Scarce Seven-spot Ladybird *Coccinella magnifica*
Five-spot Ladybird *Coccinella quinquepunctata*
Seven-spot Ladybird *Coccinella septempunctata*
11-spot Ladybird *Coccinella undecimpunctata*
Pine Ladybird *Exochomus quadripustulatus*
Orange Ladybird *Halyzia sedecimguttata*
Harlequin Ladybird *Harmonia axyridis*
Cream-streaked Ladybird *Harmonia quadripunctata*
Bryony Ladybird *Henosepilachna argus*
13-spot Ladybird *Hippodamia tredecimpunctata*
Adonis Ladybird *Hippodamia variegata*
18-spot Ladybird *Myrrha octodecimguttata*
Striped Ladybird *Myzia oblongoguttata*
14-spot Ladybird *Propylea quattuordecimpunctata*
22-spot Ladybird *Psyllobora vigintiduopunctata*
24-spot Ladybird *Subcoccinella vigintiquattuorpunctata*
16-spot Ladybird *Tytthaspis sedecimpunctata*

Calvia decemguttata (Channel Islands only)

Inconspicuous ladybirds

Horseshoe Ladybird *Clitostethus arcuatus*
Coccidula rufa
Coccidula scutellata
Hyperaspis pseudopustulata
Nephus bisignatus
Nephus quadrimaculatus
Nephus redtenbacheri
Platynaspis luteorubra
Rhyzobius chrysomeloides
Rhyzobius litura
Rhyzobius lophanthae
Scymnus auritus
Scymnus femoralis
Scymnus frontalis
Scymnus haemorrhoidalis
Scymnus interruptus
Scymnus limbatus
Scymnus nigrinus
Scymnus schmidti
Scymnus suturalis
Dot Ladybird *Stethorus pusillus*

Acknowledgements

Massive thanks must go to my fiancée, Kate, who listened to me reading each chapter (some several times as I wrote and rewrote them!) and offered constructive criticism when it was merited, as well as accepting my ever-increasing stays in the study as the final deadline approached.

Thanks must go to Julie Bailey at Bloomsbury, who asked me to write this book in the first place, and to Jenny Campbell and Susi Bailey, who worked with me to turn the manuscript into a book.

Special thanks go to Professor Helen Roy, MBE, the lead supervisor on my PhD project on the Harlequin Ladybird, who is responsible for enthusing me (and so many others) about ladybirds over the years. The rest of the ladybird research community (Peter Brown, Remy Poland, Lori Lawson Handley, Cathleen Thomas, Trish Wells and many others too numerous to mention) made me welcome for four great years before I diverted to bumblebees.

Image Credits

Bloomsbury Publishing Plc would like to thank the following for providing photographs and permission to reproduce copyright material.

While every effort has been made to trace and acknowledge all copyright holders, we would like to apologise for any errors or omissions and invite readers to inform us so that corrections can be made in any future editions of the book.

Key t = top; l = left; r = right; tl = top left; tr = top right; c = centre; cl = centre left; cr = centre right; b = bottom; bl = bottom left; br = bottom right.

AL = Alamy; FL = FLPA; G = Getty Images; IS = iStock; NPL = Nature Picture Library; RS = RSPB Images; SS = Shutterstock.

Front cover t mikroman6/G, b imageBROKER/AL; **spine** Sue Robinson/SS; **back cover** c Sally Anscombe/G, b Luís Curado/G; **1** CUTWORLD/IS; **3** Judith Borremans/NIS/Minden Pictures/G; **4** Mary C. Legg/G; **5** Mark Richardson Imaging; **6** t ARCO/NPL, b mikroman6/G; **7** bl Visuals Unlimited, Inc./Nigel Cattlin/G, br Henri Koskinen/SS; **8** t Valter Jacinto/G, b Richard Comont; **9** bl Panther Media GmbH/AL, br Richard Comont; **10** tl InsectWorld/SS, tr PHOTO FUN/SS; **11** t Anneka/SS, b Andrew Darrington/AL; **12** © Lizzie Harper; **13** bl Richard Comont, br Guenter Fischer/G; **14** Mark Moffett/Minden Pictures/G; **15** tl kris Mercer/AL, tr Richard Comont; **16** Anastasia Gapeeva/IS; **17** Giandomenico Sultano/EyeEm/G; **18** LorraineHudgins/SS; **19** Universal Images Group North America LLC/DeAgostini/AL; **20** Abrilla/SS; **22** The Natural History Museum/AL; **23** Simon Robson; **24** t Richard Comont, b wildpik/AL; **25** Brian & Sophia Fuller/AL; **26** Larry Doherty/AL; **27** t Jerry Hoare Photography/AL, b Andrew Darrington/AL; **29** t Paul Reeves Photography/SS, c Monik-a/SS, b Nigel Cattlin/AL; **30** Nature Photographers Ltd/AL; **31** t blickwinkel/AL, b thatmacroguy/SS; **32** Andia/Contributor/G; **33** Ingo Arndt/Nature Picture Library/G; **34** t Richard Comont, b Mark Richardson Imaging; **35** Richard Revels/RS; **36** t Richard Comont, b sergeklein/SS; **37** c Brett Hondow/SS, b Shane Hermans imagery/SS; **38** Patrick LORNE/Contributor/G; **39** t Eduardo Estellez/SS, b Simon Robson; **40** Andia/Contributor/G; **41** t Arterra/Contributor/G, b FLPA/Richard Becker/G; **42** c Richard Comont, b Anest/SS; **43** Losonsky/SS; **44** Westend61/G; **45** Michael Sewell/G; **46** Rick Strange/AL; **47** VEK Australia/SS; **48** Jolanda Aalbers/SS; **49** Christian Musat/SS; **50** CHROMORANGE/Dieter Möbus/AL; **51** tl InsectWorld/SS, tr Nigel Cattlin/Visuals Unlimited, Inc./G, c Tomasz Klejdyszmy/SS; **52** AlessandroZocc/IS; **53** Robert F. Sisson/Contributor/G; **54** Richard Comont; **55** dynacam/IS; **56** Protasov AN/SS; **57** Karen Grigoryan/SS; **58** yogesh_more/IS; **59** Prisma by Dukas Presseagentur GmbH/AL; **60** Dekayem/Wikipedia; **61** Neil Phillips/RS; **62** tl Muhammad Naaim/SS, tr Leschenko/IS, cl Rburi/SS, cr mb-fotos/IS; **63** InsectWorld/SS; **64** Stephen Dalton/Minden Pictures/G; **65** blickwinkel/AL; **66** t Richard Revels/RS, b Henrik Larsson/SS; **67** David Osborn/RS; **68** John Burnham/AL; **69** t Gilles San Martin, b Simon Robson; **70** l irishka5/SS, r Andrew Darrington/AL; **71** t Will Heap/G, b Africa Studio/SS; **72** Victor Suarez Naranjo/SS; **73** Adettara Photography/G; **74** Andrew Darrington/AL; **75** Richard Revels/RS; **76** Duncan Usher/AL; **77** t Gilles San Martin, b blickwinkel/AL; **78** PeterTowle/G; **79** Roger Tidman/RS; **80** t Nataba/IS, b Richard Comont; **81** Kim Taylor/NPL; **82** c Richard Comont, b Richard Comont; **83** c Gilles San Martin, b nounours/SS; **84** Daniel Borzynski/AL; **85** Lester V. Bergman/G; **87** t Nick Upton/2020VISION/NPL, b InsectWorld/SS; **88** Rod Williams/NPL; **89** Richard Comont; **90** kay roxby/SS; **92** tl Nigel Cattlin/FLPA, tr Diego Cervo/SS, b Storye book/Wikipedia; **93** t Raquel Mathias/SS, b hekakoskinen/IS; **94** Henrik Larsson/SS; **95** Sue Kennedy/RS; **96** t Sara Armas/SS, b Everett Collection, Inc./AL; **97** Libby Welch/AL; **98** t Anadolu Agency/Contributor/G, b PictureLux/The Hollywood Archive/AL; **99** t flab/AL, b eddie linssen/AL; **100** Török Dániel/Wikipedia; **101** b NASA/Wikipedia; **102** ZenShui/Eric Audras/G; **103** Anteromite/SS; **104** t Richard Comont, c Achim Mittler, Frankfurt am Main/G, b Visuals Unlimited, Inc./Nigel Cattlin/G; **105** Chris Gomersall/Nature Picture Library/G; **106** Gilles San Martin; **107** bl PHOTO FUN/SS, br Simon Robson; **108** t Mark Richardson Imaging, b Rod Williams/NPL; **109** Richard Comont; **110** bl Simon Robson, br Simon Robson; **111** Ian Rose/FLPA; **112** Nick Upton/Nature Picture Library/G; **113** ieuan/SS; **114** bl Richard Comont, br Richard Comont; **115** tl Pavel Krasensky/SS, tr Agarianna76/SS; **116** Genevieve Leaper/RS; **117** t Richard Comont, b PHOTO FUN/SS; **118** t Richard Comont, b InsectWorld/SS; **119** t Bildagentur Zoonar GmbH/SS, b Cmspic/SS; **120** aleksandr yakovlev/IS; **121** Cristian Teichner/SS; **122** bl Simon Robson, br James Lowen/FLPA; **123** Stephen Dalton/Minden Pictures/G.

Index